OARS, SAILS AND STEAM

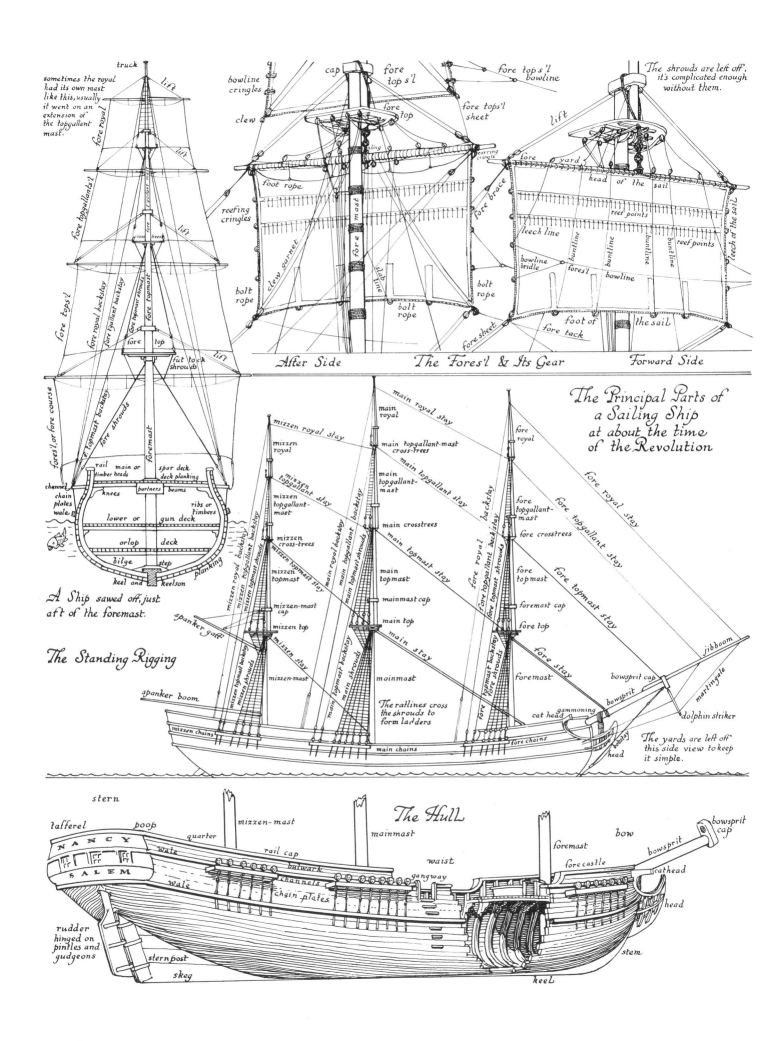

OARS, SAILS AND STEAM

A PICTURE BOOK OF SHIPS

WRITTEN AND ILLUSTRATED BY

Edwin Tunis

The Johns Hopkins University Press
Baltimore and London

Design and Typography Jos. Trautwein

Originally published in 1952 by The World Publishing Company
Johns Hopkins Paperbacks edition, 2002
2 4 6 8 9 7 5 3 1

The Johns Hopkins University Press
2715 North Charles Street
Baltimore, Maryland 21218-4363
www.press.jhu.edu

Library of Congress Cataloging-in-Publication Data
Tunis, Edwin, 1897–1973.
Oars, sails, and steam : a picture book of ships /
written and illustrated by Edwin Tunis.—
Johns Hopkins Paperbacks ed.
p.   cm.
Originally published: 1st ed. Cleveland : World Pub. Co., 1952.
ISBN 0-8018-6932-3 (pbk. : alk. paper)
1. Ships—Juvenile literature. 2. Shipbuilding—Juvenile literature.
3. Ships—Pictorial works. 4. Shipbuilding—Pictorial works. I. Title.

VM150 .T88 2002
623.8´2—dc21   2002016045

A catalog record for this book is available from the British Library.

For my nephew and godson
DAVID HUTTON
this book,
because it has in it at least
one ferryboat

ACKNOWLEDGMENTS

THIS BOOK would not have been possible without the original research of my betters who have written upon various aspects of the subject.

Grateful thanks are given to the staff of the Pratt Library in Baltimore for patient help in the pursuit of elusive facts; and to Captain Wade DeWeese, USN (ret.), Acting Director of the Museum of the U.S. Naval Academy, to Lieutenant Commander J. B. Griggs, USN, of the Academy staff, to the Navy Office of Information and Mr. Robert H. Burgess, Curator of the Mariners Museum at Newport News, for valuable assistance.

Two good friends, Dr. George E. Hardy, Jr., and Commander Elmer C. Clusman, USNR, have opened their libraries to me and allowed their brains to be picked. I thank them both.

Lib, my wife, has endured months of boats and more boats, she has typed reams about boats and has shown no sign of leaving me. I am profoundly grateful.

E. T.

The Arke Royal (1587) *Flagship of the fleet which destroyed the Spanish Armada*

ILLUSTRATIONS

OARS, SAILS AND STEAM

HIS BOOK follows the main changes in boats from the very beginning up to the present. It doesn't begin to tell all there is to learn about boats; it doesn't even begin to show all the kinds of boats which have existed; but it does show the interesting and important types, and it tries to explain why they are interesting and why they were developed.

It is hard to be sure of anything in connection with very ancient boats. No one then realized that, later, men would be interested in how those boats were built and how they were operated, so a crude drawing and a phrase or two is usually all we know about them.

It took hundreds of years for man to learn the use of an oar. Years passed in thousands before he could let go of that oar and depend on the wind alone. Learning to sail was a cut-and-try process; no science and little experiment was applied to it. There is still no reason to believe that sail had reached the end of its possibilities when steam put it out of business. Who knows what the windjammer might become if her problems were studied by modern methods and science? But sail is gone and there's many a fellow at sea who can thank his stars that he doesn't have to wrestle at midnight with half a ton of wet, slapping canvas on an icy yard arm. Steam itself may now be superseded by forces which are barely beginning to be understood.

There are a few "seagoing" terms in this book, but the first time each one is used, it is explained in the margin of the page. At the back of the book there is a list which explains the meanings of many more of these terms.

The diagrams in the front of the book will help in finding and understanding the parts of boats.

OF COURSE nobody really knows; still it's a good guess that the first boat of all was just a log which fell into a stream. A man could straddle it and move it a little, by paddling with his hands, but he would get along better by pushing against the bottom of the stream with a stick. The first time a man did this was almost too long ago to be imagined.

IT SEEMS the first step towards a real boat would be to hollow out the log. That would make it lighter, easier to handle and much drier; and you could carry things in it which wouldn't stay "put" on top of a log. These boats are called dugouts; and they are still made and used, right here in the United States.

Coracle   *Kayak*   *Birchbark Canoe*

LOGS BIG enough to be hollowed into dugouts weren't found everywhere. In many places people had to learn to build boats with whatever they had. Though the ancient Britons made big dugouts, they also built coracles, which were really very large baskets covered with skins. The Eskimos made their tricky kayaks of skins stretched over a bone frame. The northern American Indians learned to sew the bark of the yellow birch tree on a light wood frame for their canoes, which are still unbeatable on small streams. From the Susquehanna River southward, the Indians, having no birch, made dugouts.

Egyptian Boat Model (c. 8000 B.C.)

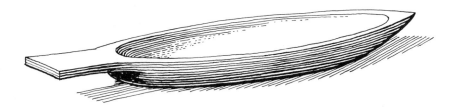

THIS ISN'T a real boat; it's a clay model of a boat found in a tomb at Fayoum in Egypt. However, there must have been boats which looked like this at the time the model was put in the tomb. Scientists say that was as much as ten thousand years ago.

It's hard to tell how big the original of the Fayoum boat was, but you can guess she was not very big. The Egyptians still use a skiff called a nugger which looks a lot like the model.

A skiff is a boat small enough for one man to handle.

Man Sailing in Fayoum Boat

Egyptian Nugger

Ambatch

Because wood was scarce the Egyptians learned to make boats by lashing together bundles of reeds. Far up the Nile they still make one that way, which they call an ambatch. The old ones were finished off in long points which stuck up quite high above the water, at the bow and stern. The ambatch is that way at the bow, but it doesn't seem to have any stern at all.

Probably the Fayoum boat was paddled; it's doubtful that she sailed, though her owner may have stood up and spread out his coat once in a while, to let the wind push him along. It's likely that's how sailing was discovered.

The bow is the front end of a boat and the stern is the back end.

13

Egyptian Sailboat (c. 4700 B.C.)

The hull is the whole body of a boat. It is what floats.

A mast is an upright "stick" for supporting sails and ropes.

Braces are ropes to hold the top of a sail against the wind. Those which hold it back at the bottom are called sheets.

Backstays and forestays pull against each other from the top of a mast to stiffen it.

SAILING WAS discovered a long time ago. There is a crude drawing of a sailboat on an Egyptian vase made about eight thousand years ago. You can make out that it was meant to show a sail but that's about all.

Let's jump a little matter of thirteen hundred years. Then the wall paintings really tell something about the boats they represent. This one at the top of the page is a combination of them. Because the hull was built of reeds, like the ambatch, the mast had to be two-legged, so as to spread its weight. A single-pole mast would punch right through the bottom of a reed boat.

This boat would sail ahead of the wind and a little to either side, but as soon as the wind was crosswise the sail was lowered and the paddlers bent their backs. It's pretty certain also that they didn't sail when the wind blew too hard. That fellow sitting on the roof, hanging on to the braces, would have his hands full in any kind of a breeze!

There are things about this boat that stayed nearly the same for many a century. The yard across the mast to hold the square sail, the backstays, the forestay, the braces and the sheets are all in place as they were on the *Lightning* in 1854.

Why it took two and sometimes three men to steer a craft like this you'll have to figure out for yourself.

14

Egyptian Oared Ship (c. 1600 B.C.)

HERE IS a vessel large enough to be called a ship. It was about sixty-five feet long. The Egyptians were now building their ships of wood, which they may have bought from the Phoenicians.

Perhaps the oddest thing about this ship is the heavy rope truss, to stiffen her lengthwise. This is a good sign there was no keel to stiffen her. The old way of building reed boats with high bows and sterns was still followed with wood.

More than the hull was improved: the sail could be raised and lowered. Each of the yards is made of two pieces lapped over at the middle and lashed. This was partly to make use of the natural taper of the trees.

The loose ropes looped across the sail were for the upper yard to hang from when it was lowered. The braces are fastened near the helmsman; there's too much boat here for a man to manage the braces in his hands. It's doubtful that she ever sailed across the wind, she would make great leeway with no keel and the mast is not properly braced to take the strain.

There are two rudders made like big oars lashed over the sides. The helmsmen steered by rotating the rudder with the tiller, which you can see projecting from the top of the oar.

At first any large boat was called a ship. Later the meaning of ship changed somewhat.

The long timber on the outside of a boat's bottom is her keel. An added one on the inside is a keelson.

Leeway is slipping sidewise away from the wind. The lee side is the one towards which the wind is blowing.

15

Phoenician Warboat and Merchantman (750 B.C.)

THE Phoenicians were high-class traders and sailors and they worked at both for three thousand years. They sailed from Tyre at the eastern end of the Mediterranean all the way to England for tin; they sailed down the west coast of Africa and, possibly, clear around the Cape of Good Hope and up the other coast. They were great at trade but not so good at drawing, so it's hard to be sure exactly what their ships looked like. The two drawn here are just an honest try.

It is certain that the metal ram or beak, which had been invented in Crete, appeared on their ships some twenty-six hundred years ago. They had rowers who sat exposed to the weather, they had square sails with no lower yard, and there was an eye painted on each side of the bow, so the ship could "see" where she was going. Chinese junks still have these eyes. The Phoenicians seem to have thought of a ship as a horse. They used carved heads of horses as figureheads, at the bows of their merchant ships.

Greek War Galley (250 B.C.)

THIS IS a Greek warship looking for a scrap, twenty-two hundred years ago. The Greeks had another word for it, but we call a large boat with many oars a galley. When there's only one bank of oars, a galley is a *unireme,* when it has two banks, like this one, it's a *bireme.*

Galleys were built long and narrow, for speed and easy rowing. Almost all vessels which were built to be rowed were "long" ships, and they were principally war boats, but not always. Two main maneuvers were used in battle: first, a head-on ramming attack intended to sink an opponent, or at the least, to confuse him badly; second, boarding and slugging it out, sword-to-helmet.

A Greek war galley had a sail, as you can see. It was probably used only when the wind was fair, or nearly so, because sailing across the wind would heel the boat over and make a mess of rowing.

The "tail feathers" which show at the sterns of Greek boats seem to have been extensions of the sternpost and the long timbers of the hull. This must have been for ornament, since there doesn't appear to be any boat-building reason for it.

There's a ladder projecting behind the galley's stern. This shows in the old Greek vase paintings. It may have been used for getting aboard another ship, or it may have been used to get up the mast.

A boat heels over when wind pressure makes it lean to one side.

The sternpost is the upright timber in the middle of the stern.

Greek Merchantman (250 B.C.)

NOT ALL Greek ships were galleys. This cargo boat was painted on the same vase with the war boat on page 17 so it must have been afloat at the same time. It isn't a "long ship," it's a "round ship." Speed was no object, but carrying capacity was; also, though it could be rowed when the wind failed, it was really a sailboat.

The mast was not only stayed from bow to stern, as earlier ships had been, but it also had ropes, called shrouds, from its top to the *sides* of the boat. Wind pressure from the side must have been expected, so the vessel could and probably did sail across the wind and perhaps could even angle a little bit into it, with the yard braced sharp around and the windward clew pulled forward.

Those loose loops of rope, hanging down around the sails of these Greek boats, are brails, used to gather up the sail to the yard. One end was fastened to the yard; then the rope was carried under the bottom of the sail from behind, up across the front of the sail, over the yard and down to the deck. When the deck ends were all pulled at the same time, the sail was raised, something like a Venetian blind.

The "fence" along the side of the ship was simply to keep sea water from splashing on the cargo.

The clews are the lower corners of a square sail. The windward is the side from which the wind is blowing.

Roman Trireme (c. 200 B.C.)

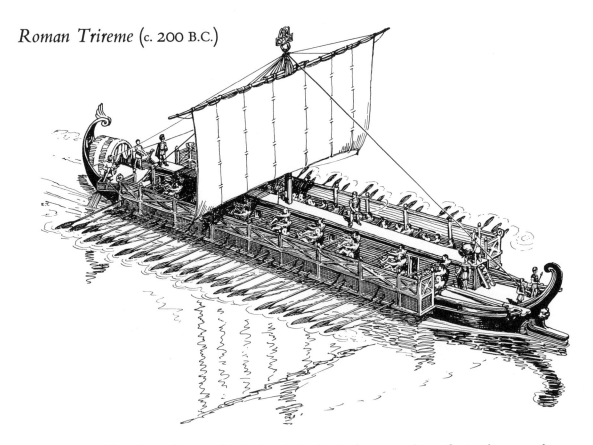

THE Romans took to the sea because they had to do it; they had no natural taste for it. They started by copying a Carthaginian galley which they found stranded. The rowers were trained in frames set up on the shore.

Romans were best at hand-to-hand fighting, so they invented the *corvus*. This was a kind of gangplank, with hooks at its outer end, for grappling an enemy ship and boarding it. It was highly successful; in time Romans spoke of the Mediterranean as "our sea." Slaves for rowing were plentiful, so there was no serious loss when an enemy galley sheared off a whole bank of oars, reducing the rowers to dog meat.

The rowers worked in rhythm, as oarsmen in a racing shell do now. A drum on deck beat the stroke for them, stepping up the beat when more speed was needed. Overseers, equipped with persuaders, patrolled a platform above the rowers, encouraging them to keep the stroke and put their backs into it.

The ship in the large drawing is a trireme, with oars on three levels. The sketch at the bottom of this page shows one way to arrange the rowers for this. The oar-holes were not in the hull but in balconies hung outside the hull. This gave a rower the advantage of a long handle inboard. Galleys were steered like the Egyptian ships, by rudder-oars on both sides.

Anything inboard is inside the rail of a boat.

Sketch of Trireme Oar Arrangement

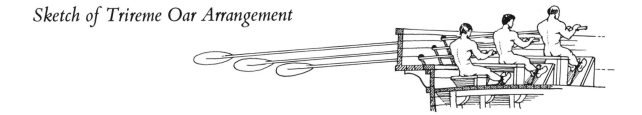

Roman Shroud Rigging

end of lanyard

shroud siezing

wooden deadeye

lanyard to tighten shroud

wooden deadeye

bronze fitting on hull

The quarter is the part of the hull just forward of the stern.

Rigging includes all the ropes and spars to hold the sails and manage them.

To tack is to sail a zigzag course against the wind.

Buntlines are ropes attached to the lower edge of a square sail for brailing it up to the yard.

Once the Mediterranean became "our sea," Romans naturally went into commerce. Possibly they copied the round ships of the Greeks; certainly they had round ships and left us very clear pictures of them in their carvings. These ships had low, round bows and high, round sterns, with the sternpost almost invariably ending in a carved swan. The two rudders were operated from projections built onto the hull at the quarters.

These were quite sizable boats. From descriptions we know that some were as much as a hundred feet long. The Romans improved rigging. They tightened their shrouds with deadeyes. The small drawing will explain how deadeyes work.

A little sail was added above the mainsail to catch more wind and a short mast was slanted forward over the bow to hold still another sail, called the *artemon*, which made the ship steer better. The Romans still could not tack into the wind. Caesar was astonished when he saw it done by the Venetii in Gaul.

Rome invented footropes, which were strung behind the yard for the sailors to stand on while they worked with the sail. Later these disappeared and then were invented all over again in the north of Europe. Roman ships had odd buntlines which ran under strips of canvas sewed to the forward face of the sail.

The Oseberg Ship—Norse (835 A.D.)

Construction of Norse Ship

FOR THE next sure knowledge of ships after Roman times, we have to skip centuries and go far north to Norway. But here are ships we really know about, because the actual hulls themselves have been dug out of Norse graves and preserved.

They are wonderful boats, light, strong and almost perfectly shaped for sliding through the water. Not until whaleboats were built in the eighteen hundreds were boats again as well designed. The Norse ships were clinker-built, that is, each plank of the hull lapped over the plank below it; and each plank was perfectly shaped to fit the curves of the hull. Little cleats were hewn on the inside face of each plank and these were lashed tightly to the ribs. Some islanders in the Pacific build boats exactly the same way.

There were holes in the upper planks of the Norse ships for oars and some had little covers to close the holes while the ship was sailing. The shields along the sides probably served to keep the crew dry, for there was very little freeboard and only their perfect shape kept the ships from swamping. They were given plenty of chance to swamp. The Norsemen sailed them all up and down the coast of Europe and across to North America.

Don't get the idea they were only skiffs. The Oseberg ship is seventy-five feet long and sixteen feet wide. It had one square sail and only one rudder. The rudder hung over the right side which, because of this, was called the steer-board side, hence the right side of a ship is the starboard side, right down to now.

Freeboard is the height of the side from the waterline.

21

The Mora *of William the Conqueror* (1066)

THIS DRAWING shows all we know of ships at the time the Normans conquered England. Duke William used the *Mora* to cross the Channel. There may be some doubt about how much the drawing is like the Duke's boat, because it is taken from the Bayeaux Tapestry which was woven as much as a hundred years after the invasion and there is no certainty that the weavers knew much about ships.

You can see what a tub the *Mora* was, compared to the Norse ships, but you can also see that this ship is a relative of theirs. It has the high stem and sternpost, the one rudder over the starboard side, the shields along the rail and the one square sail on a mast in the center of the ship. Ships in northern waters were getting rounder and they continued to do so.

The stem is the upright timber at the front end of the keel.

There was never any idea that the *Mora* could fight another ship. She merely carried troops. If she came close enough to an enemy vessel, some arrows would be shot and there might be a boarding party, but the ships themselves would not attempt to damage one another.

The *Mora* is estimated to have been some sixty-five feet long. The sternpost decoration seems to have marked her as the flagship; at least none of the other ships in the Tapestry have it. The lantern at the masthead was carried only on the *Mora*, so that the rest of the fleet could follow her at night.

English Esnecca, a Warship, Time of Henry III (1250)

THIS WAS called a warship and was used as one, but it wasn't built for the job. When he saw a fight developing, King Henry simply hired or took merchant ships and altered them for war. A ship was a movable fort. Temporary "castles" were built at the bow and stern and a smaller "top castle" was put at the masthead. The name of the quarters forward for the crew in all ships is still the "fo'c'sle"; and there is still a "top" on all ships with square sails.

This ship was even tubbier than the *Mora* but she was a better sailer. Her rigging had all the gear for sail handling that the Romans had, plus two things the Romans never knew they needed: first, bowlines, for pulling the windward side of a sail forward, so as to sail nearer the wind; second, a bonnet, a strip of canvas laced to the bottom of the sail. It could be taken off to make the sail smaller when the wind blew too hard.

King Henry's ship still has a rudder over the starboard side, but before this, about the year 1200, a few ships had rudders hung in the middle, back of the sternpost. Old records speak of two kinds of ships, cogs and nefs. Today it isn't certain what the difference between them was, but perhaps the nef was steered with a side rudder, the cog with a stern rudder.

Mediterranean ships at this time still looked quite Roman, with low bows and high sterns, but they had two definite masts now, rigged with triangular lateen sails on very long yards. The Southerners tightened their shrouds with tackle rather than with deadeyes.

The later top is a platform at the head of the lowest section of a mast.

Rope rigged between pulleys to increase pull is called tackle. Deadeyes work the same, but have only holes in wooden blocks instead of pulleys.

23

English Ship with Sternpost Rudder (1340)

THIS IS the first ship which surely shows a stern rudder. It was steered with a tiller and the man who moved the tiller had to be told what to do, because he was under the after castle and couldn't see where the ship was headed. Even so, a ship was easier to sail with a stern rudder because she turned more quickly.

When a ship is careened she is pulled over on her side.

This was a true war vessel, the castles were part of the ship, not just temporary platforms. These ships all had heavy timbers applied to the outer surface of their hulls. The long ones were called wales; the short, vertical ones were skids. They took the wear and strain when the ship was careened ashore to have her bottom cleaned. They were also a protection in case of collision.

This ship had a bowsprit projecting forward on the starboard side of the stem, but it carried no sails. The bowlines were led out to it, yet its prime purpose was the same as the Roman *corvus*. It was a boarding bridge, and that is not an anchor hanging at its tip; it's a grappling hook, to secure it to an enemy vessel. Another new feature is the ratlines. Ratlines are light lines, evenly spaced across the shrouds to form ladders for getting up the mast.

It should be mentioned here that the first cannon were mounted on ships by the French in 1356. The fight was on! The French began using three masts also. To use two, it was necessary to move the mast one way or the other, but the main mast had always been in the middle of the ship, it would never do to move it, so, to make the ship balance, they had to have three!

24

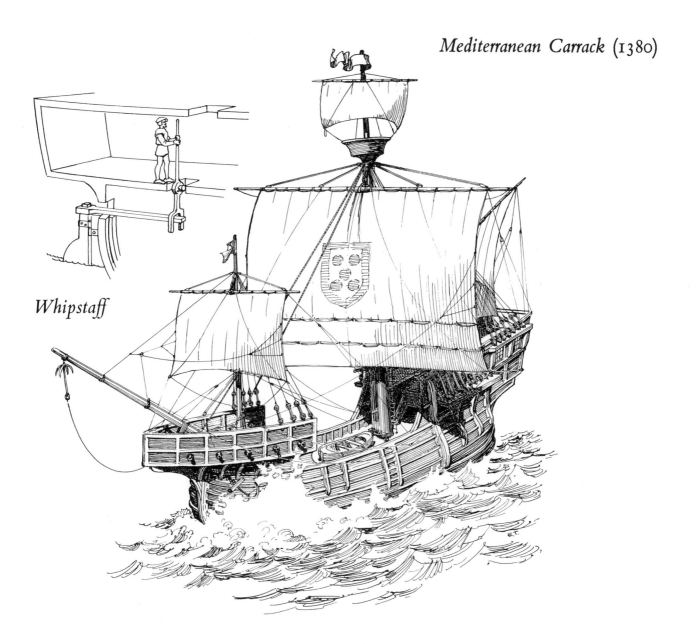

Whipstaff

WITH THE improvement of ships the Northerners began taking wool and salt fish to the Mediterranean ports; and the Southerners began bringing spices and silks to Southampton and to Bruges. Each looked over the other's ships and what he liked, he swiped. In the south the first carracks appeared with two masts. The Northerners quickly copied the hull, but stuck to their own sail arrangement. Just as quickly, the Southerners caught on to the sailing advantages of three masts, so they, too, began to use three, and added another sail above the main course.

Now the ships were pretty much alike, but in the south, where the sea was calm, the long platform over the bow—the chief mark of a carrack—was lower than in the north. The bowsprit remained to starboard of the stem (it stayed there until 1670) and was designed for boarding, though more rigging began to be attached to it. The hull developed a new quirk which persisted for a long time. It was built with tumble home: the ship was made narrower at the level of the main deck than at the waterline. This was supposed to make it hard for an enemy to board.

Masts were still made from a single pole, but they were getting much too tall for this to be practical. Steering became a little easier with the invention of the whipstaff, to move the tiller from the deck above it.

If there is more than one sail on a mast, the lowest one is called the course.

25

Flemish Carrack (c. 1450)

Anything in the middle of the ship is amidships, not amidship.

SHE WAS called a *kraeck,* but that was just a Flemish way of saying carrack. Her likeness to her southern sister is easy to see—it is the differences which are interesting. She was planned for rougher weather than the southerner. She is nearly decked over amidships to keep, as the fellow said, the water out of the cellar. How'd you like to stand on that deck in a heavy sea?

It was because of rough water that she was built so high forward and had shorter masts than the southern ship. For the same reason she had no tops'l as the southerner had and her castles were framed to be covered for protection from rain and cold.

The tops'l is the second sail from the deck. It is set on the topmast, above the top.

The galleries were balconies, open or enclosed with windows.

Port: The left side; also an opening in the hull.

There are a couple of other points about this carrack; she still had shields on her rail, though they were only ornaments, and she showed the beginnings of the stern and quarter galleries which were prominent features of ships for three hundred years afterwards. The entry port, low on the port quarter, is very unusual. She doesn't show any heavy guns, but she must have been planned for fighting; those are ammunition lifts hanging from the tops. Those necklaces on the masts are parrels. They steadied the yards while allowing them to be shifted around. All the essential gear of a full-rigged ship is here. It is said that rigging developed so completely in the first fifty years of the fifteenth century (1400–1450) that a sailor of that time would be entirely at home on the *Constitution* in 1814.

26

The Santa Maria *of Columbus* (1492)

ONE MORE carrack and a famous one—the *Santa Maria*. She has been called a ship, a caravel and a galleon but she was really a carrack. More bad models have been made of her than of any other vessel that ever floated, even including the *Mayflower*. There probably never was a picture of the *Santa Maria* drawn by anyone who had actually seen her, so it's impossible to know exactly what she looked like. The drawing above is a Spanish carrack of Columbus' time and so may come fairly close to her appearance.

It's a little hard to pin down the meaning of the word "ship," because it has meant different things at different times. Once it was any large vessel; then it became a vessel with a "head" at her bow (of which more presently); after that it described the way a boat was rigged, three masts with square sails; and nowadays it is once more any large vessel!

Before we go on to ships and heads, we should remember the two other craft which went with Columbus. One of them brought him home. The *Nina* and the *Pinta* were both caravels. Smaller than a carrack but somewhat the same shape, they had no bowsprit and were not decked over. Most caravels had triangular lateen sails, but the *Pinta* had been given square sails before she left Palos. The *Nina* started with lateens, but the Admiral had her re-rigged square, at the Azores. Skippers of later days would have approved; for a long haul, the square-rig is better than the fore-and-aft.

When sails stand parallel to the keel instead of crosswise the rig is fore-and-aft.

27

The Great Harry (1515)

HER REAL name was as fancy as she was: the *Henry Grace à Dieu.* She was the wonder of her time and the first ship of the English navy. Henry VII built her and she served his descendants down to Elizabeth.

The *Great Harry* was a true ship, not because of her rig but because of her head. Whether the bow-platform of the carrack was moved down to become the head, or the ram of the galley (it was added to some carracks) was moved up, is something for the hairsplitters to argue about. The fact remains that the head developed after the carrack and remained in various forms for a long, long time. It was a platform at first, but it projected around the stem and in front of it, not over it, as the carrack bow was built.

There were new things about the *Harry.* She had four masts: fore, main, mizzen and bona-venture. Her masts were not of one piece but of three, locked together and rigged to stiffness with taut ropes. Except forward, she bristled with guns. As to castles, you can see she had "real" ones!

Managing the *Great Harry* at sea must have been like sailing a warehouse and a very high, top-heavy one at that. They tricked her out with golden sails and she ferried King Henry VIII across the Channel for his powwow with the King of France at the Field of the Cloth of Gold— but he resolutely refused to come home in her!

28

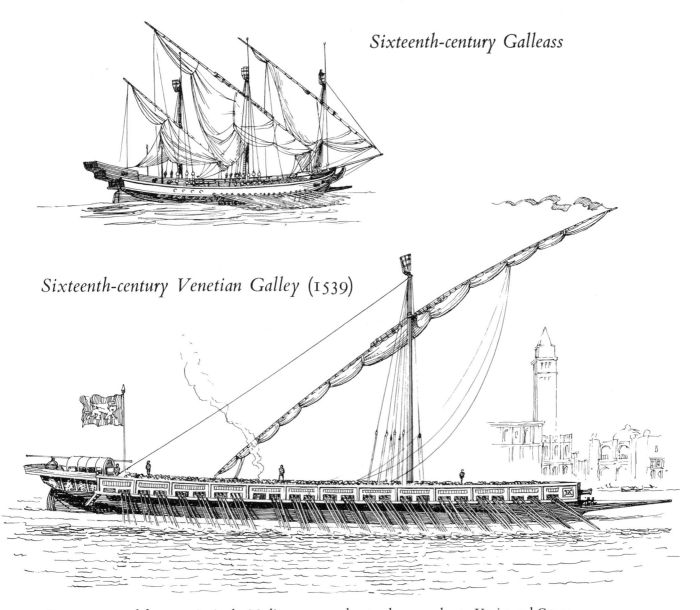

Sixteenth-century Galleass

Sixteenth-century Venetian Galley (1539)

GALLEYS operated for centuries in the Mediterranean and not only as war boats. Venice and Genoa ran regular packet lines of them to Alexandria, Constantinople and way stations. Long before Eli Whitney, the Venetians used standard, interchangeable fittings for their boats and kept spare parts and maintenance crews at all ports of call. By the way, the smoke in the drawing is from the cooking fire.

There had to be slaves to row galleys, so, as war captives became scarcer, convicts took up rowing. France operated galleys rowed by convicts up to 1748; chiefly as a means of disposing of the convicts.

In a fight there wasn't much a galley could do except make one head-on charge at the enemy. If it worked it was perfect; but long, narrow boats don't turn quickly. A sailing vessel with guns could hit and turn and hit again, while the galley was getting squared away to ram.

Admirals as a class hate to change things. So, instead of forgetting about oars, they developed a contraption called a galleass. It had oars, a ram at the bow and three sails instead of just one. It was shorter, wider and higher than a galley and it carried guns, especially heavy ones to shoot forward. This was the only real advantage the galleass had. Ships at this time had such high heads that they couldn't mount guns to shoot forward. The one advantage wasn't enough. Galleasses were still too long to handle easily and their height made the oars too long for good rowing.

29

Elizabethan Galleon (1587)

THE TIMES of Queen Elizabeth were great times for men and great times for ships. The ship on this page was built only about seventy years after the *Great Harry*, but look at the difference! The changes are not in principle but in refinement.

The three-decker in the above drawing is a galleon from the same fleet. Galleons were the handiest fighting ships then afloat. She looks gaudy and she was gaudy; her poop was high and narrow and her carved, projecting head made it impossible to shoot forward, but the underwater part of her hull was fish-shaped—streamlined we'd call it. The masts had additional bracing from backstays, led from the masthead to the bulwarks. The rigging, too, had been improved. That sail under the bowsprit is the sprits'l. It served the same purpose as the Roman *artemon*—it balanced the pressure of the wind on the high stern.

By the way, the tall ships that the Spanish used for bringing home gold and spice have come to be called galleons; in fact, most people call any ship with a high poop a galleon, but the original type had three decks and a rather low bow, and was of moderate size. These could sail anywhere in the world and they very nearly did.

The poop is the high deck at the stern, the place of honor.

30

Elizabethan Crumster (c. 1590)

WHILE THE galleons were sailing around the world, somebody had to stay home and work. A busy trade went on in nearby waters—up, down and across the Channel. Home products and Spanish loot were peddled to Europe and shipped from one English port to another. All this was the business of the sturdy little crumster.

Fat-bodied, to carry as much as possible, mounting a few light guns, because you never knew what you might meet at sea or when you might have to turn-to and fight for queen and country, the crumster was about half the size of a galleon, with only two decks and two masts, but with all kinds of sails.

The mainmast carried not only the ordinary, square course and tops'l, but had, in addition, a long sprit to which a fore-and-aft sail could be rigged. This is another kind of sprits'l. The crumster had the regular kind also, hung under her bowsprit. The mizzenmast was short and set only a lateen.

On the forestay a new sail appeared, called, reasonably enough, the forestays'l. It was triangular and was the first of the heads'ls, which later included the jib, the flying jib and others. It was also the first of the stays'ls, which later were set on stays between the masts.

The crumster had no need for a fighting top, so the topmast shrouds were spread by two arms, called crosstrees.

Early Seventeenth-century English Merchantman (1607–1632)

This is the kind of ship in which the first English settlers in America crossed the Atlantic. It would be nice to draw the ship and give it a name, but it would be a fake. Actually nobody knows exactly what any one of these ships looked like. They didn't become famous until a couple of hundred years later; in their own time nobody thought to make pictures of them.

The bilge is the curve between the side of the hull and the bottom. It is also the unpleasant liquid which collects inside below the curve.

They were all ordinary merchant ships and all pretty much like this drawing. It will serve for the *Sarah Constant*, the *Goodspeed* or the *Discovery*. Perhaps you haven't heard of them. They brought the Virginia settlers to Jamestown in 1607. In 1620 the *Mayflower* of the Puritans would have looked a lot like this; and in 1632 the *Arke* of the Marylanders would not have been noticeably different.

Traveling in any of them was no picnic. They ran with rats and reeked of bilge. In good weather passengers did their own cooking at an open fire on deck. In bad weather they were kept under the deck. They did no cooking then and probably had very little interest in it anyway.

32

The Sovereign of the Seas (1637)

PHINEAS PETT built her for King Charles I and she was the fanciest ship that ever *was* built! From head to stern gallery she was covered with carving painted red and gold. One fifth of her whole cost was for decoration; and her whole cost was a pretty penny, so much that the king levied a special tax to help pay for her. That tax was one of the reasons he lost his head.

In spite of this trouble the people loved the *Sovereign* and they firmly said, "No!" when Oliver Cromwell suggested painting her black.

This vessel was heavily armed and a good fighting ship, but always a cranky sailer because she was top-heavy. Late in life she was razeed, that is made lower, by having one deck removed. She was the first Ship of the Line to be rigged with three masts instead of four. In order to spread the same amount of canvas on one less mast, she was given a sail above the topgallants'l. It was called the royal (not the royal sail) probably because it was first set on the flagstaff provided for the Royal Standard.

A Ship of the Line was one large and powerful enough to fight in the first line of battle.

The gans'l or topgallants'l is above the tops'l; it's the third sail from the deck.

33

ABOUT THE time King Charles II took his giddy whirl at the English throne, fighting ships began to be rated according to the number of their guns. The Fourth Rate in this drawing had fifty guns and two gun decks.

Navies need small boats as well as large ones. The little fellow here is a ketch, described as being, "to goe to and fro and to carry messages between shipp and shipp."

Carving, which disappeared under the Roundheads, came back with Charles, but it no longer ran wild and was confined mostly to the stern and quarter galleries and to the head, which now became much smaller. These ships still had raised quarter-decks and forecastles, but by the trick of using an open rail for these higher parts, the vessel was given a sweep of sheer line from stem to stern. The poop deck was left off entirely. The round tuck stern by which British craft came to be identified at a glance first appeared on Restoration ships; and on them the bowsprit was first set over the stem, instead of to one side of it. Reef points, to shorten a sail by tying a section up to the yard, began to be used along with bonnets.

The silly-looking little mast at the tip of the bowsprit, for the sprits'l tops'l, now became standard equipment. A few stays'ls began to be set between the masts. The tops became simple platforms. Steering was still by the ancient whipstaff and very difficult with large ships. In 1711 the ninety-gun Second Rate *Ossory* became the first ship to be built with a steering wheel.

In Elizabeth's time the quarter-deck was forward of and lower than the poop. Now, with the poop gone, it ran all the way to the stern.

Sheer is the dip of the top line of the hull.

Eighteenth-century English Hoy

THIS WAS a work boat, as the earlier crumster was. Hoys were usually small. They fished and hauled coal, sailed in all weather and did a little smuggling on the side.

That big gaff sail probably took most of the wind from the square main course. It is worth noting that revenue cutters, which chased hoys in the way of business, used the same rig with the square course left off.

The little Dutch yacht below has the gaff sail with only a square tops'l. Both boats carry two headsails though, where the crumster had only one. The yacht was used by the Dutch Admiralty about 1750. Her side-cabin windows are framed in carving, to look as much like quarter galleries as possible.

Dutch waters are shallow, so the Hollanders invented lee boards, swung over the sides to use in place of a keel. These kept the boat from making leeway, yet they could swing back and drag easily over a shoal.

The peak or top of a fore-and-aft sail is fastened to a spar called a gaff. If there is no boom at the bottom it is a gaff sail, or loose-footed sail.

Dutch Admiralty Yacht (1750)

35

H.M.S. Victory (1765)

A strake is one line of hull planking. The sheer strake is at the top of the bulwark —the rail, that is.

THIS SHIP is still kept, fully rigged, at the Portsmouth Navy Yard in England. Her tall sides, some thirty feet from waterline to sheer strake, were the famous "wooden walls of England," which His Majesty's subjects once comfortably believed to be impregnable. The *Victory* fought against us in the Revolution and was Nelson's flagship at Trafalgar.

The design really hadn't changed much from the time of Charles II. The stern and quarter still had galleries. The head changed, but only a little. The sheer line was a trifle straighter than that of Charles' Fourth Rate, but the *Victory* went back to the high poop deck, which earlier ships had abandoned. This was removed late in her career. The *Victory* is 186 feet long; she carries 108 guns and hence is a First Rate.

Ships had been built in America almost from the beginning, but only European types. After 1700 Americans were developing their own designs. The Marylanders began to tinker with clippers and in New England the Marblehead heeltapper appeared. (See drawing on page 77.)

The heeltapper was primarily a fishing boat; small but high-pooped, with round "apple bows." That far she was old-fashioned; it was her rig that was new. She had no square sails at all. She wore the complete boom-and-gaff rig, just about as it appears on small working schooners today.

British East Indiaman (c. 1775)

THE British East India Company was a government in itself. It raised its own armies and maintained a great fleet of ships to bring home its goods from the East. These vessels sailed to India the hard way, around the Cape of Good Hope. Using the dependable trade winds, they voyaged across the Indian Ocean for days on end, carrying full sets of stuns'ls, two sprits'ls and a sprits'l tops'l; and shortening sail only at night. They all mounted guns and were a little old-fashioned in the way of fancy heads and galleries.

The French developed the smart little lugger, which they still use in their coastal waters. These boats were sharper than had been customary; that is, they were narrower and deeper for their length and their bows were shaped to cut through the water, rather than to push it aside. The deep keel added to their fore-'n-aft rig made them fine sailers when close-hauled.

Maybe the lugsail was "designed" by chopping off the lower end of a lateen. On the French lugger, it must have been a mean sail to handle; having to get the foot of the fores'l over the main stay every time the boat came about would be a nuisance. The long, level bowsprit and the equally long, level after-bumkin seem to belong to this rig; luggers still have them that way.

Studding sails, or stuns'ls are extra sails for use in light air. They are set on booms extending from the ends of the yards.

Close-hauled: sailing as nearly into the wind as possible.

To change from one tack to the other is to go about.

The *American Continental Frigate* Raleigh (1776)

WHEN THE American colonies decided to fight for independence, they had no navy of any kind. The Congress commissioned private ships to fight and ordered thirteen frigates to be built. One of these was the *Raleigh*. She was built at Portsmouth, New Hampshire, in the record time of sixty days! When she put to sea she was quickly captured. The British admired her so much that they "took off her lines." These drawings the Admiralty still has and from them we know what the *Raleigh* looked like.

What was a frigate? First it was a small galley; then any ship with a flush deck; next, in the *Raleigh's* time, it was a ship of thirty-two to forty-four guns (in the British Navy, a Fifth Rate).

The drawing on page 76 gives a fairly good view of the deck. You can see the raised fo'c'sle and the guns behind their ports; and you can get an idea of the rigging of the shrouds and back-stays. On page 20 the chains and deadeyes are enlarged to make them clearer.

Around the base of each mast is its fife rail, to which the halyards were belayed. Braces, sheets and tacks were belayed at the pin rails along the bulwarks. On top of the bulwarks are the hammock nettings, where all sleeping hammocks were stowed when the ship cleared for action; to get them out of the way and to help protect the gunners.

A halyard is a rope for raising a sail.

To belay a rope is to fasten its end.

The bulwarks are the solid rails along the ship's sides at the main deck.

38

Twenty-four-pound Truck Gun (1776)

HERE'S A closer look at one of the *Raleigh's* truck guns, called that because it was mounted on wheels called trucks. These guns were made of cast iron, in various sizes, named from the weight of the round shot which fitted them. By comparing the numbers on the drawing with the list below you can find the names of the main parts of the gun. This gun is not being served by a full crew.

1. barrel	6. cascable	10. train tackle	14. sponge
2. breech	7. carriage	11. outhaul tackle	15. match
3. muzzle	8. quoin	12. breeching	16. water bucket
4. touchhole	9. truck	13. rammer	17. powder box
5. trunnion			

The gun is inboard in loading position, but still so close to the bulwark that the rammer handle has to project through the gunport. The Gun Captain sticks a wire into the touchhole; the cartridge (a woolen bag of powder) and a wad are rammed back into the barrel until the Captain feels the cartridge touch his wire. While the shot is being rammed in, the Gunner pokes a hole in the cartridge with his wire, pulls the wire out and pours a little loose powder into the touchhole.

At a signal the train tackle is released and the outhaul tackles are used to roll the gun forward, until its muzzle projects about two feet from the gunport. The Gun Captain touches the burning match to the touchhole. There is a smoky flash, a loud noise, fire flies up through the touchhole and an iron ball is thrown from the muzzle about twelve hundred yards seaward.

As the shot goes out the gun ramps back and takes up short against the breeching. The train tackle is hauled taut, to keep the gun back; and the Gunner puts his thumb on the touchhole to kill sparks. The barrel is swabbed out with a wet sheepskin "sponge," to clean it and to make sure all sparks are dead in the bore. The gun is ready to load again. Time, with a good crew: less than a minute.

The U.S. Frigate Constitution (1797)

AT THE end of the Revolution the Navy had only two small ships left. So, as soon as the dust had settled, Congress set about replacements. The first three new vessels were oversize frigates: the *United States*, forty-four guns, built at Philadelphia and disrespectfully known as the "Old Wagon," from her lumbering pace; the *Constitution*, forty-four guns, built at Boston and known as "Old Ironsides," for reasons no American needs to have explained; and the *Constellation*, thirty-eight guns, built at Baltimore. All three were designed by Joshua Humphreys. Some have said the three were of French design, because the French also built large frigates, but this is nonsense.

The *Constitution* was a good ship under sail, but it was her strength, rather than her speed, that made her formidable. Humphreys did away with most of the traditional frippery. Her stern was squared across and the head was greatly reduced. There was no raised quarter-deck, the spar deck made a full sweep from stem to stern. She was and is painted black, with a broad white band marking the line of her gunports.

The yard, once used for the sprits'l, is still under the bowsprit, but not for a sail; it serves only to spread the stays which stiffen the jib boom. In her heyday she carried a tremendous spread of canvas; her great weight made it possible and also made it necessary. She can still carry it. She has been rebuilt and is a regularly comissioned Navy ship.

The open, upper or main deck is often called the spar deck.

The jib boom is a spar extending from the bowsprit. The foot of the jib is secured to it.

The North River Steamboat of Clermont (1807)

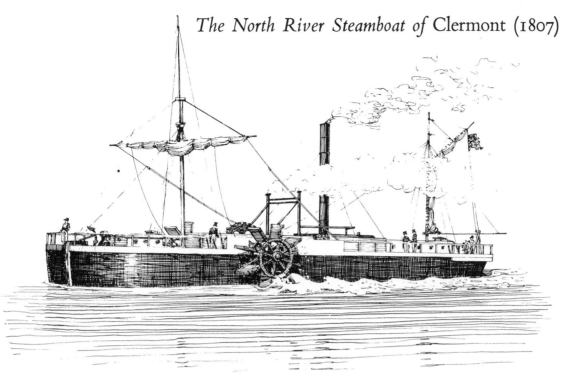

ROBERT FULTON didn't invent the steamboat. No one man did, but Fulton built a practical one, and that started their general use.

Jonathan Hulls patented a steam-driven boat in England in 1736; and in 1787 John Fitch operated the contraption at the bottom of this page on the Delaware River, at a speed of two miles per hour. The *Clermont* made five miles an hour against the current of the Hudson and was hailed as a miracle by all the people who had gathered in the fearful hope of seeing her blow up.

Boiler explosions were common in the early days of steam, and it became customary for timid ladies to ride in Safety Barges, towed at the end of a long hawser, behind the steamboat.

The name of Colonel John Stevens should not be forgotten. He experimented with the screw propeller and built the paddle steamer *Phoenix* at New York, the same year Fulton built the *Clermont*. Quite possibly the *Phoenix* was the better boat, but she was completed a little too late and Fulton got not only the glory, but the sole right to operate steamers on all New York inland waters. Stevens took his boat to Philadelphia, where she worked for her living for some years. The trip from Sandy Hook to Cape Henlopen made the *Phoenix* the first steamboat to go to sea.

The screw propeller is used on most boats. It works in the water the way the thread on a bolt works in a nut.

Fitch's Steamboat (1787)

Baltimore Clipper Schooner (1814)

THE Baltimore clippers are not to be confused with the clipper ships which developed later. The Baltimore design seems to have grown out of a type used in Bermuda and the West Indies; "very much on the Bermuda mould," says an old Baltimore newspaper advertisement.

The clippers were superbly fast sailers, far faster than anything afloat in their time. For this reason they were much used as privateers in the War of 1812 and later, for the same reason, as slavers.

During that second war against England, the American Government issued "letters of marque" to privately owned vessels, which then had the privilege of annoying the British. The clippers did this with great success and with considerable profits to their owners. Profits came from the sale of "prizes," that is, captured ships and their cargoes. The *Rossie*, commanded by doughty Joshua Barney, sank or captured fifteen British ships in forty-five days. Captain Boyle, in the *Chasseur*, declared a one-ship blockade of the British Isles!

These Baltimore clippers were very sharp—long, narrow and deep and the keel grew deeper as it ran aft, giving them what is called "drag," but not in the sense of slowing them down. The bow was shaped so that it was thin below the waterline and tended to slice through the water.

A tops'l schooner was fore-and-aft rigged except for one or more square tops'ls.

Most of these boats had two masts and were rigged as tops'l schooners, though there were departures from this. Their deep keels made them "stiff" in the water, so they could carry much more sail than an ordinary hull of their size could stand.

The Dutch

Steam Naval Packet Curaçao (1827)

SAIL AND steam were developing together. Steam was new and sail had not yet reached its peak; there was little in the way of competition for some years.

The little *Curaçao* was built on the Thames in 1826 as an auxiliary steamer and was sold to the Dutch Navy. She had two separate engines, one for each paddle wheel and a full set of fore-and-aft sails. She was sent across to the island for which she was named, in 1827, using her engines for a large part of the voyage. She returned uneventfully and made the trip several more times in the course of her career.

An auxiliary is a boat in which sails are helped by power.

The *Curaçao* was certainly the first steamship to cross the Atlantic from east to west. Let's look at the record and see if she doesn't have a fair claim as first in the other direction also!

The American ship *Savannah* is usually credited with being the first steam vessel to cross. Cross she did in 1819, but—she was a sailing ship. An engine and paddle wheels were put in her and she was sent to England, to be sold as a steamer to the Czar of Russia. It took her twenty-seven days to cross; that's six hundred and forty-eight hours; and she ran her engines a total of eighty of those hours. She was almost out of fuel halfway over, so the skipper saved enough to get up steam and make a grand entrance into Liverpool Harbor. The Czar turned her down.

The Savannah (1819)

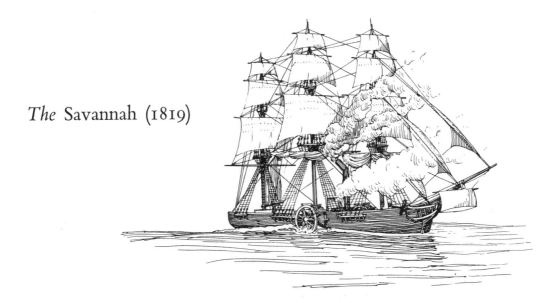

The Baltimore Clipper Ship Ann McKim (1836)

IT WAS only a question of time before someone would build a large Baltimore-clipper hull, give it three masts, and rig it as a ship. Isaac McKim did it, and the *Ann McKim* was the result. To us now she looks pretty much like any other clipper ship, but to the people who had seen only blunt-nosed hookers, like the *Savannah* (without her paddles), the *Ann McKim* was as new as a rocket airliner would be today.

A ship in this sense is a vessel with square sails on three masts.

No money was spared on her. Brains and experience went into her design and pride and workmanship into her timbers. She was wonderful! Sail? She eased-along in a dead calm and hardly shortened her canvas for a gale.

Isaac put her in the "China trade," around the Horn and across the Pacific. She worked at it until 1847 and behaved like a lady, but she had one fault—so narrow was her cleaver-shaped hull that she had too little space for cargo.

So, in a sense she was a failure, but McKim had shown the way and authorities all over the world salute the little *Ann McKim* as the grandmother of all clipper ships. Master shipbuilders found it easy to modify her extreme lines and produce the greatest sailing vessels of all time, the Yankee Clippers of the Fifties.

The S.S. Great Western (1837)

WE Americans tried new ways of building sailing ships and no one ever built them better; we also pioneered in steam, but to tell the plain truth, we weren't so good at steamboats. We appear not to have had much real love for them at first. "Stove on a raft!" we snorted in disgust.

It was the English who took to steam. Their early paddle boats were so beautiful that their lines are still followed for fine power yachts. Beside them the flat-topped, high-sided Yankee steamers looked clumsy and "mechanical."

The *Great Western* was English and the first ship built to carry out a scheduled run across the Atlantic. That is, she was designed to carry enough coal for steaming all the way across. This was not "impossible" as most people said it was, but it left her sadly short on cargo space.

She used her sails when they would help, but she kept her engines turning all the way across. She made ninety crossings in her life. Her best time was twelve days and fourteen hours, far better than any of the sailing packets could do. In this drawing she is shown passing a New England fishing pink at sea.

The honor of being the first ship to cross entirely under steam goes to another English boat, the *Sirius*. She started three days ahead of the *Great Western* and reached New York with only a couple of hours lead. She ran out of coal off Sandy Hook, so her skipper had all her spars chopped up and burned to keep up steam and bring her in.

There was a Great Eastern too but she was an outsize freak. She laid the Atlantic cable.

45

The U.S. Steam Frigate Mississippi (1840)

THIS WAS the third steam vessel built for the United States Navy. The first two were both named *Fulton. Fulton I* blew up in 1829. *Fulton II* was largely experimental, but she infected Matthew Calbraith Perry with the idea that steam was the thing for the Navy. Matthew was Oliver Hazard's brother and a good man, too. He battled Congress and the whole Navy, including the enlisted men, until he got a real steam frigate, which was the *Mississippi*.

She had sails of course—not even Perry would have thought of taking a ship to sea without them. Her mainmast was shoved aft a little to make room for the stack, otherwise her rig was the same as any other frigate. Sails were wisdom too; a round shot in the paddle box would cripple a steamer and you never knew when the connecting rod would break. She must have depended mainly on her engines, because she ordinarily carried no yards "across" above the tops'ls.

The connecting rod transferred power from the engine to the paddle wheels.

The *Mississippi* proved to be quite a girl. Perry used her as his flagship in the Mexican War and took her along when he went to the Orient to negotiate the United States' first treaty with Japan. She was also the first steamship to go all the way around the world. Her career ended at Norfolk in 1863 when she was burnt to keep her out of the hands of the Confederates.

A Black Ball Line Packet Ship (1850)

It HAD never been customary for sailing ships to have a fixed time of departure. The ship sailed when the tide was right, when she had a full cargo or when the cook recovered from a binge; perhaps tomorrow, perhaps a week from Friday. Shippers never hurried about getting their goods to the wharf either, they knew the ship would wait for them. All this, of course, annoyed the passengers but there was nothing they could do, except wait and accumulate soiled clothes.

Steamers hadn't yet given the sailing packets any trouble. They were faster, but only a little faster and they cost so much to run that their passenger and freight rates were sky-high. But these came down gradually and speed increased. The pressure went on slowly and the windjammers had to improve or quit.

A packet is a ship which keeps to a regular run.

The leader in improvement was the Black Ball Line. William Webb built them a fleet of fine ships, designed for good speed but with plenty of cargo room and with full attention to the comfort of the passengers. Best of all they sailed on a definite schedule.

With these innovations the sailing packets managed to compete with steam for some thirty years. As the going got tougher, so did the crews. Expense had to be kept down to stay in the race, so lower and lower grade seamen were hired and the time came when "packet rat" was as low as a sailor could get.

American Whaling Bark (c. 1850)

WHALERS HAD no schedules whatever. They went where the whales were, or where the skipper hoped they were—to the South Atlantic, to the south coast of Australia, to the Bering Sea. They killed whales, chained the carcasses to the ship's side, cut off the blubber in chunks, melted it down in the try-works on deck and ran off the oil into barrels, which they stowed in the hold.

There's not space here to talk about the breath-stopping job of catching a whale; that takes a whole book. If you're interested try *The Cruise of the Cachelot* and of course *Moby Dick*.

Whaling vessels weren't much different from ordinary cargo boats. They were almost always three-masted, but more often bark-rigged than ship-rigged. They always had plenty of beam and were thoroughly seaworthy. Speed was no object—some voyages lasted two years or more.

It was her special gear, not her build or her rig, which identified a whaler. There were five or six beautiful cedar whaleboats secured to davits and resting on brackets along her bulwarks, with spare boats stowed on deck. There was a wide opening in the bulwarks, where the cutting platform was hung. On deck, just aft of the foremast, were the brick try-works and high on the fore and main masts were the lookout stages, from which rang the famous cry, "Blo-o-ow!"

A bark has square sails on the fore and main masts but fore-and-aft on the mizzen.

Davits are small derricks for raising small boats from the water.

Whaleboat

Lookout Stage

The Paddle Coaster Union (1851)

ALONG THE Atlantic seaboard, steamers began to compete with the coastwise sailing vessels and while he was building windjamming packets and clippers, William Webb also turned a hand to the construction of these wooden paddle coasters. The *Union* is one of them which he built.

Steam river traffic developed rapidly too. Using sails on narrow, twisting rivers was never very practical because of puffy winds caused by the nearness of the land. The steamboat was the right answer and, as a result, was important in opening up the remoter parts of the country.

In the East boats with keels and some draft could be used, but these wouldn't serve for the shallow Mississippi and Missouri. What was needed was a boat that could "run on a heavy dew." Captain Henry Shreeve solved the problem with flat-bottomed steamers.

The draft of a boat is the depth of water needed to float her.

Usually these were driven by stern-wheels, which could be raised in very shallow water. They had suspended gangplanks at the bow, so they could run right up to the bank, for there were no wharves. They burned wood which was plentiful. Each line owned woodlots along the river where its boats stopped to pick up fuel.

Mississippi River Stern-wheeler

The Schooner Yacht America (1851)

As THIS is written, it is just a hundred years since the *America's* cup was won, off the Isle of Wight. All through that century British and Canadian yachts have tried vainly, time and again, to recapture "The Old Mug." Sir Thomas Lipton made so many tries that, even on this side of the water, many hoped his fine sportsmanship might finally be rewarded. Probably it didn't matter too much to him, for the sport was in the race, not in the trophy.

The *America* was the project of the Stevens brothers and their friends. The Stevenses were the sons of Colonel John Stevens who built the steamer *Phoenix*.

There is a famous and oft-told tale of the original great race. When the first returning yacht was sighted, Queen Victoria asked a man with a spyglass what boat it was. "The *America*, Your Majesty," he replied. "Who is second?" asked the Queen. "Your Majesty, there is no second!"

The Clipper Ship Lightning (1854)

MOST LONG hauls were still made in sail, without much attention to how long it took to get there. The discovery of gold in California and Australia introduced the need for speed which resulted in the clippers, and proved their qualities. Men and supplies were wanted at distant points, and quickly, and Yankee clippers made wonderful runs around the Horn to Frisco.

When Jimmy Baines of Liverpool wanted to start sailings to Australia, he found no English ships fast enough for the job. So he went to Boston and ordered a fleet from Donald McKay, the master shipbuilder of them all.

The *Lightning* was not only the fastest of that fleet, she was the fastest. Her record was never equaled in sail. On March 1, 1854, the Old Man wrote in the logbook, "Wind South, strong gales . . 18 to 18½ knots per hour, lee rail under . . distance run in 24 hours 436 miles." Thirty years passed before a steamship did better. The *Lightning* was *always* fast too: sixty-four days Australia to England, the best previous time being seventy-five days.

This ship was 226 feet long at the waterline and had the sharpest ends of any clipper ever built. She carried a huge spread of sail, 13,000 square yards. That much sail needed strong rigging and the *Lightning* had it; her backstays were of Russian hemp, more than three and a half inches thick.

The Chinese Junk

THE JUNK is put here because the Yankee Clipper sailors were the first Americans to become really familiar with it. Perhaps the word "junk" has given us a bad impression of these vessels; if so, it's a wrong impression. They are fine craft, sturdily built, seaworthy and wonderful sailers. Each batten in the heavy lug sail has a separate sheet, so the top of the sail, as well as its bottom, can be close-hauled to make a junk sail nearer the wind than any other kind of sailboat.

The battens are the cross-poles which take the place of yards.

There are dozens of types, each with its own home port. Some have eyes at the bow, a trick which came from Egypt centuries ago by way of the trading Arabs. The general shape suggests the lines of a Mediterranean carrack, but the junk goes further back than that.

The ship in the drawing is a northern trader, usually called a "red-headed junk" from its bright red stern. This stern has a slot in which the big rudder can be moved up and down; when it is down it serves the purpose of a keel, in keeping the boat from making leeway.

The stem is raked 'way forward, so the nose can be easily put ashore in shallow water. The bottom is perfectly flat; a junk left high and dry stands neatly upright. By the way, junks have always been divided into watertight compartments; this is a fairly new idea in the West.

52

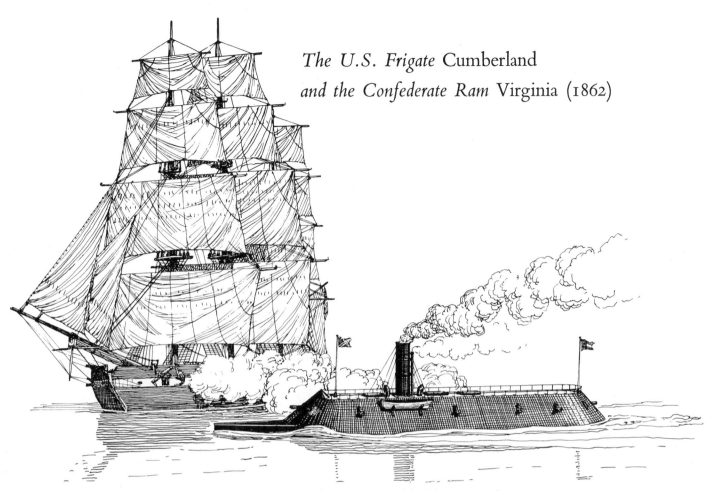

The U.S. Frigate Cumberland
and the Confederate Ram Virginia (1862)

THE Navy has never cared to change very fast. Though it started building steam frigates in 1840, it did not stop building sailing frigates. There came a time when sail and steam met and clashed. When the smoke cleared, sail had come off a bad second.

The Civil War was on. The U.S. Frigate *Cumberland* was lying in Hampton Roads. Out of the Elizabeth River came the new Confederate ironclad ram *Virginia*, which not long before had been the U.S. Steam Frigate *Merrimac*. Now she had been stripped of her masts and her thick oak timbers covered with iron plates; as clumsy a craft as ever floated, but to a wooden sailing vessel she was deadly.

There was no wind, and the *Cumberland* couldn't budge. Her new shell-firing Dahlgren guns hardly dented the "Iron Barn"; the *Virginia* just sat there and chewed her to pieces, then rammed her, and went on to polish off the *Congress* before retiring for the day.

Next day the *Virginia* met something of her own sort, only more so. This was the new U.S. ironclad *Monitor* and this time the story was different. John Ericsson designed the *Monitor*. She was flat and hard to hit, but the thing that made her deadly was her revolving turret. There were only two guns in it and they couldn't be both fired at once, but they could be fired in any direction, independently of the ship. That did it; monitors were in! Naval design was revolutionized twice in two days.

U.S. *Ironclad* Monitor (1862)

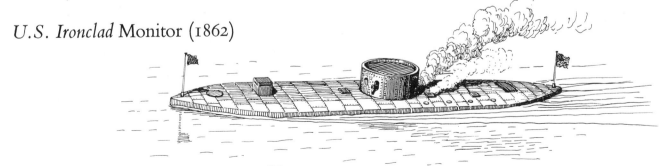

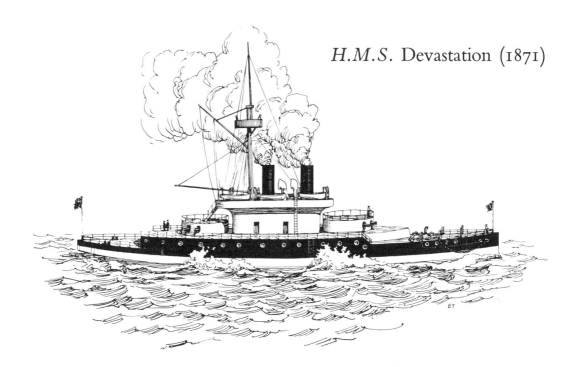

H.M.S. Devastation (1871)

AFTER THE War between the States, the American Navy went to sleep. It developed the monitor idea no further, but the British Navy did. The English tried many experimental boats (one of which turned over and drowned three hundred men) before they built the *Devastation*. She had two turrets and an odd, narrow deckhouse, which allowed the guns to be turned nearly all the way around a circle. She depended almost entirely on the four guns in the turrets, because for a while it was believed that nothing more, except a ram bow, was needed.

The *Devastation* was a screw steamer. The British Navy settled the argument as to which was better by hitching a side-wheeler and a propeller boat, of equal power, stern to stern and opening the throttles. The paddle boat stirred up more foam, but she was pulled backwards a couple of miles an hour.

At about this time the Americans hit on the idea of using merchant vessels as reserve warships. At the bottom of this page is the Pacific Mail Line iron screw steamer *City of Pekin* which was armed and, as she kept her schedule between San Francisco and Yokohama, was rated a Navy cruiser. She was built in 1874 and was actually taken over by the Navy in 1898, but was used only as a troop transport. The *City of Pekin* did her part in helping to extinguish the clippers by covering her run across the Pacific in sixteen days.

S.S. City of Pekin (1874)

Barkentine (c. 1880)

STEAMBOATS needed fewer hands than sailing ships and so became cheaper to run. To stay in business, the owners of sailing vessels tried to lighten the work on their boats. One of the first things they did was to divide the big tops'ls into two, the upper and the lower, the way this old windjammer carries them. She is a barkentine, with only the foremast square-rigged.

Fore-and-aft sails needed fewer men, so square sails began disappearing until, in 1951, there are only two commercial square-riggers working in the world. For a while, around 1900, very large fore-and-aft rigged ships were tried. They used steam winches to raise and lower sails and had five and six masts; one had seven.

BARK   BRIG   BRIGANTINE

SCHOONER   TOPS'L SCHOONER   SLOOP

In addition to the barkentine and to full-rigged ships, like the *Lightning* these are the most usual rigs of later sailing vessels:

Bark: Like the whaler, three-masted, but with no square sails on the mizzenmast.

Brig: A two-masted square-rigger.

Brigantine: Two-masted, with square sails on the foremast only.

Schooner: Two or more masts, all rigged fore-and-aft. Developed in America and now used all over the world, more than any other sailing rig.

Tops'l schooner: All masts rigged fore-and-aft, except for a square tops'l and usually a top gallants'l, on the foremast. This rig has practically disappeared.

Sloop: One mast, fore-and-aft rigged. Long ago sloops had square tops'ls.

The U.S.S. Chicago (1885)

THE English copied the *Monitor* and the Americans sat tight. When the United States was ready to go again, it copied the English. The result was the "Squadron of Evolution," sometimes called the "White Squadron," of which the *Chicago* was the first. You'll notice our old shellbacks still couldn't imagine a battleship without sails, even if they were never used. This idea was so fixed that the Navy actually built the *Chesapeake* in 1900, so as to have a sailing ship for training midshipmen who, when trained, would serve only in steamships!

The "Squadron of Evolution" evolved after five years or so into the *Maine* and her sisters. Even these clung to some rigging for imaginary sails. The *Maine* was blown up in Havana harbor, an incident which helped to start the Spanish-American War. The others, built at about the same time, and some older ships like the *Chicago* gave good accounts of themselves in that war.

Like the *Devastation*, all these ships were given ram bows. This was because Navy men everywhere were impressed by some fancy ramming done at the Battle of Lissa, between the Austrians and the Italians, in 1866.

The U.S.S. Maine (1890)

Tramp Steamer (c. 1900)

THESE TOUGH old steamers gradually took over the work of the clippers. As their diesel-driven, modern sisters do yet, they went anywhere and everywhere that cargo could be carried. They seldom knew when they reached a port where new orders would send them. Like the whalers, they were sometimes gone from home waters for a couple of years.

The peculiar design was for the handling of cargo. The two lower deck levels had very big hatches with sectional covers which could be removed to permit loading and unloading. Usually goods were handled by tackle attached to the cargo booms on the two masts.

The pilothouse and the officers' quarters were in the high center section. The crew lived in the fo'c'sle. At sea in heavy weather, when a tramp was laden and "down in the water," seas frequently broke clear over the cargo decks.

Big bridges and tunnels have cut the number of ferryboats in use, but once they were almost the only way in and out of New York City and were absolutely necessary to many towns. Day in and day out they waddled back and forth, always pulling out just in time to be missed.

Robert Fulton invented double-ended ferries, as well as the slip for docking them and the derrick-raised gangplank which can be set flush with the deck.

Double-ended Paddle Ferry

The Great S.S. Mauretania (1908)

Aside from one in a rope, knot is a nautical mile. It equals 6,075 feet.

THE *Mauretania*. There's a new ship with that name now, but she can never hope to live up to it, fine vessel though she be. When the old *Mauretania* was built she was far ahead of any liner which had ever floated. She was larger, faster and more luxurious. It took five years' work to build her. She made a handy twenty-five knots and, on her maiden voyage, captured what the English call "the Blue Ribband of the Atlantic." She held it more than twenty years, while the Germans and French nearly burst their boilers trying to take it away from her. She crossed in four days, ten hours and forty-one minutes; few today can equal it.

The warship wasn't doing badly either. At last the admirals decided it was no disgrace to put to sea without the gear for sails. The *Washington* was launched in 1906. She kept the ram on her bow but, fortunately, she never had to use it, since she would probably have done herself as much damage as she did her opponent. Aside from that she had quite a businesslike, modern look, more so than the more powerful basket-mast battlewagons which were built some years later.

The U.S. Armored Cruiser Washington (1906)

Gloucester Fishing Schooner (1910)

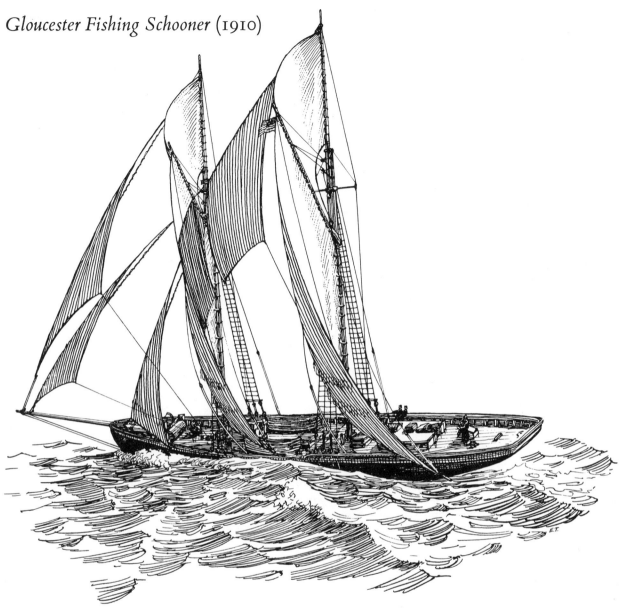

ALL SCHOONERS have two or more masts and have fore-and-aft sails, though in old days they could carry a square tops'l or two and still be schooners. The Baltimore clippers were usually rigged that way.

Quite different and very fine schooners turned up in New England: the Gloucester Fishermen. These worked on the Grand Banks before the days of refrigeration. When the catch was in the holds, the boats staged a race for home. The winning skipper, since he was the likeliest to arrive with fresh fish, got the best price.

As new boats were built they were made sharper and sharper. Yacht designers were called in to mold as much speed into them as possible. Gradually an interest grew in racing them for the sake of the race, regardless of the price of fish. This interest came to be shared by the public and kept the schooners in business long after power-driven, refrigerated boats had taken over the real work on the banks. In fact the last of those craft were really racing boats. Being too large to operate profitably as fishermen, their losses were paid by wealthy sportsmen.

The drawing is of one of the older, working schooners which were fast, but fast, for practical purposes only.

U.S. Torpedo-boat Destroyer, Camouflaged (1918)

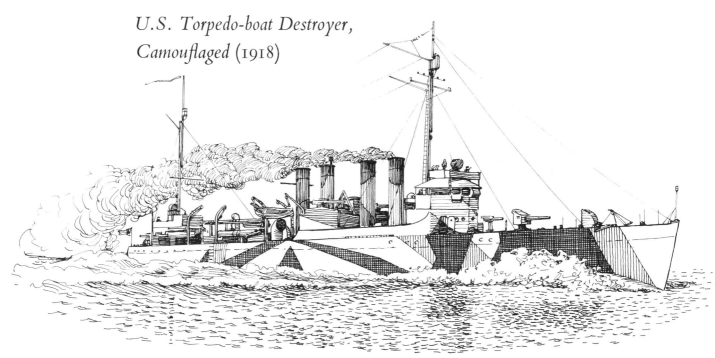

In World War I luxury liners became troop transports and cargo vessels carried mountains of supplies to Europe, but without the destroyers they would all have had to stay home. Germany had good submarines with good crews, and the combination had taken charge of the Atlantic and left the British Navy up a creek.

Late in the War the submarines were largely bottled up by the North Sea Mine Barrage, but the first effective answer to them was the revival of the old convoy system, which the Dutch had used to protect their East Indiamen in the seventeenth century. In World War I a convoy meant a large fleet of slow merchant vessels surrounded and protected by fast destroyers. These last were the "Four Stackers," long, narrow, low and very, very wet. They pitched like mustangs, rolled half over, traveled at thirty knots and dealt with the subs. They bracketed the pigboats with ash cans (depth charges) and punched them with torpedoes and gunfire. It's unlikely that this War could have been won without the destroyers.

Another type of boat which did pretty well against the undersea menace was the little Sub Chaser. Half as fast as a destroyer, they were, if possible, twice as uncomfortable. They were only a hundred and ten feet long and too slow for convoy duty, but patrolling from bases in groups of three, they accounted for many a pigboat.

U.S. Submarine Chaser (1918)

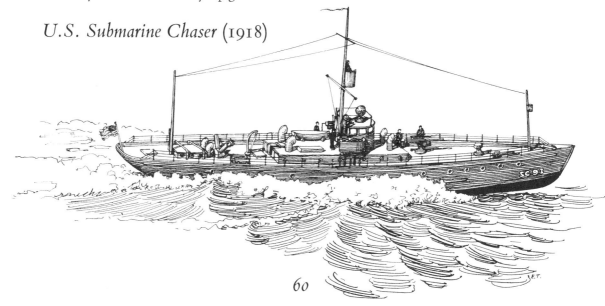

The U.S.S. Idaho (1919)

ALTHOUGH THE big American battlewagons saw very little action in World War I, nobody suggested abandoning the type. The *Idaho* was commissioned at about the end of that scrap. In this drawing she is shown as she was originally built, with the cage masts which were the latest thing at the time. The idea was that shells could pass clear through the masts without bringing them down. It was a good idea, but the masts made the ship top-heavy, so they were abandoned.

The *Idaho* had hers replaced with modern tripod masts in the early thirties and, as an elderly lady, gave a magnificent account of herself at Saipan, Iwo Jima and Okinawa. She was damaged at the last named place but she showed up in Sagami Bay with the Occupation Fleet.

World War II produced some new kinds of boats, all motor-driven. The landing ships for tanks (LST's), and the landing craft for infantry (LCI's), are built to run right up on an enemy beach. The little Patrol Torpedo boats (PT boats), Mosquitoes they're called, are plenty fast.

Newest Patrol Torpedo Boat (1951)

Landing Ship Tank and Landing Craft Infantry (1944)

The Queen Mary (1936)

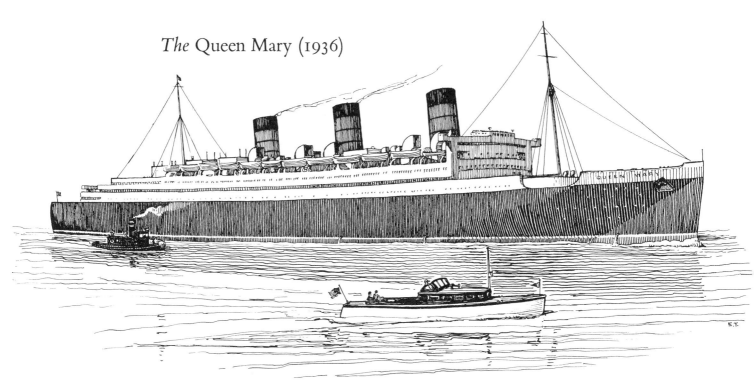

THE French *Normandie* was larger, and the *Queen Elizabeth* is too, but neither of them ever beat the *Mary*. Her record crossing of the Atlantic in three days, twenty hours and forty-two minutes has stood since 1938.

She was built to dominate the Atlantic in both luxury and speed. Nothing in the way of such fancy trimmings as swimming pools, nurseries and dog kennels was left out of her.

During World War II the *Queen Mary* was stripped of her elegant tinsel and as the *Mauretania* did in World War I, she went to work as a troopship. She carried men by thousands. Now she has been cleaned up and refurbished and is back on the old run, for a while anyway.

Hitler made his subs harder to lick by running them in "wolf packs," but this time the convoys had the added protection of planes flown from small Escort Carriers, usually converted from cargo vessels. On the surface, instead of the big Four Stackers, small Destroyer Escorts were used. These were powered by four big diesels and armed with depth bombs, torpedoes and deck guns. They could be built much faster and cheaper than destroyers.

Convoy with Destroyer Escort and Escort Carrier (1943)

"J" Class Sloop Yacht Ranger (1937)

THE YACHT *America* was a schooner (she's on page 50), but the cup she won has been defended for many years by sloops. Racing committees have drawn up a succession of complicated rules which have continually altered the design of racing sailboats and have produced the "J" Class sloop as the largest type. One of these, *Ranger,* made the latest defense of the *America's* Cup in 1937 against the English *Endeavour.*

The differences between one "J" and another are not noticeable to the average eye. Like all racers, they carry elaborate sets of perfectly cut sails, which are changed to meet every condition of the wind and the race. The masthead is more than a hundred feet from the water and handling the huge sails requires a very large crew. These sloops are not for pleasure cruising; they are built for racing and for little else.

Cup Defender Sloop Vigilant (1893)

63

U.S. *Aircraft Carrier* Saratoga (1925)

The funnel is the smokestack and the bridge is the high deck from which the ship is controlled.

IN 1911 Eugene Ely landed an airplane on a platform added to the afterdeck of a battleship. In 1922 the experimental carrier *Langley* was contrived by attaching a wooden flight deck to the hull of an old collier. When a plane landed on that deck, a hook on its tail-skid gathered up ropes which were stretched across the deck between sand-bags. As it accumulated more and more weight, the plane gradually slowed down and usually stopped before it had run off the far end of the deck. However, the *Langley* provided opportunity for experiment and study.

In 1925 the battle cruiser *Saratoga* was rebuilt into the Navy's first line-of-battle carrier, with almost two acres of flight decks. To an old sailor she was a strange-looking craft. Barnacled admirals shuddered at the sight of her funnel and bridge shoved clear over on the starboard side. Later carriers have both together as a single "island." On later carriers, too, the flight deck is full width the whole length of the ship, and elevators for bringing up planes operate outside the ship's hull. Small boats are stowed in recesses in the hull and anti-aircraft guns are tucked under the hatch of the flight deck.

A single section of the hangar decks of one of the newer carriers, such as the *Midway,* is big enough for a football game. Here on the hangar decks, engines are warmed up and the planes are delivered "top-side" ready for immediate flight.

A carrier has a double crew: one to handle planes and another to operate and fight the ship. Both jobs are tough. Keeping the decks cleared, so that one plane doesn't land on another, is an oversized traffic problem; and maneuvering the ship so that she will always head into the wind when planes are landing or taking off is an equally large problem of another sort. Then, too, there are guns to be manned.

Age didn't keep the *Saratoga* from fighting. She was torpedoed off Pearl Harbor and again in the Battle of the Solomons. At Iwo Jima in 1945, she was hit by seven kamakazes. In between she threw punches at Bougainville, Rabaul, Wotje, Eniwetok and Surabaya. Many a man shed tears for the *Sara,* when they made her a sitting duck for the atom bomb.

"Liberty" Ship (1943)

Up to now, modern wars have been won by "logistics." This seems to mean getting men where they're needed and supplying them with what it takes to live and fight. Not battleships, nor carriers, nor submarines will do that. It requires cargo ships.

When we have to fight a war, we build far more freighters than any other kind of ship, and we build them fast. In World War II strange new methods were used to finish them quickly. Hulls were built upside down; and the plates were welded instead of being riveted. Deckhouses were completed ashore and lifted aboard in one piece by derricks.

More than one type of ship was built; some were called "Liberty" ships and some "Victory" ships, but they were all turned out like sausages and went overboard in faster time than the little *Raleigh* was built in the Revolution.

Oil Tanker

Special ships are used for certain kinds of cargo. Oil, for instance, requires tankers. In these nearly the whole hull is divided into a series of tanks, with the engines pushed as far aft as possible.

The same idea is used in the odd-looking whalebacks which carry grain on the Great Lakes, but in a whaleback the hold is not divided at all. Most of the ship is one big grain bin. A whaleback can be loaded in two hours, by simply pouring the grain into her; and unloaded with clamshell buckets in half a day. These ships are disappearing.

The hold of a ship is the space for carrying cargo.

Great Lakes Whaleback

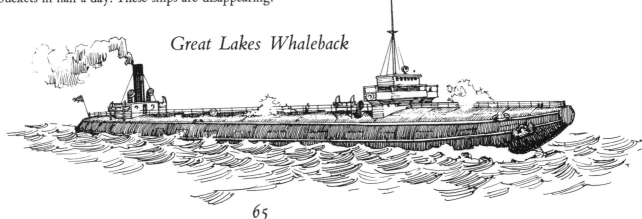

65

The U.S.S. Missouri (1951)

THOUGH THE *Missouri* put in some embarrassing days on a very public mudbank, it is the signing of the Japanese surrender on her deck which will be remembered. The iron mess table which was used that day is preserved at the U. S. Naval Academy in Annapolis. That was September 2, 1945.

She was completed late in the War, but not too late to be at Iwo Jima and at Okinawa, first with Task Force 58 and later as Admiral Halsey's flagship. In 1948 the "Big Mo" was the only battleship in commission in the Navy. Recently, lying in deep water off the coast of Korea, she has done a fine job of lobbing tons of steel onto military targets some twenty miles inland.

The war boat has come a long way from the trireme and her guns have come almost as far from the *Raleigh's* twenty-four-pounders, and not only in size. The *Raleigh's* gunners waited for their vessel to swing broadside to the enemy and hopefully cut loose as the roll of the ship swung the gun muzzles upward. The *Missouri's* gunners can set instruments which spot the target within a yard or so, allow for wind pressure, the roll of the ship, the curve of the shell path and other difficult calculations; they feed information to the gun and the gun puts a sixteen-inch shell on an invisible target.

66

THE FIRST American submarine, called the *Marine Turtle*, was designed and built by David Bushnell in 1776. Her crew was one overworked man who, in absolute darkness, unable to see where he was going, propelled the "ship" by turning a crank. The *Turtle* was not entirely successful, but one night she threw an awful scare into the crew of H.M.S. *Eagle!*

Robert Fulton did a lot of experimenting with *sailing* submarines and with steam-driven ones. Actually underwater boats were hopeless until two things had been invented: the internal combustion engine and the periscope.

By World War I great advances had been made and by World War II many subs were sizable ships. As they grew, they accumulated more and more deck gear, in the way of bridges and guns, until they looked like small battleships. Now this is all cleared away. Submarines have as few lumps on the hull as possible and they slip through the water with little resistance.

The Dutch invented the snorkel which is now being used by the U.S. Navy. It stays at the surface while the submarine is submerged. It is so small that it can hardly be seen, but air can be sucked down to the ship from it; air for the crew to breathe and air to run the diesel engines.

A periscope is a long tube with mirrors in it. Looking into the lower end, a submariner can see his surroundings at the surface.

The S.S. United States (1951)

THIS NEW liner is the largest vessel ever built in America and one of the largest in the world. Most ships are built on shore and are launched by being slid into the water. This one was built in a dry-dock and, when her hull was finished, was simply floated by flooding the dock. The upper part of the 990-foot hull was actually 30 feet longer than the dock, but the keel was short enough to allow a tight fit. She is an air-conditioned luxury liner, capable of carrying two thousand passengers, and she has already surpassed in speed the record of the *Queen Mary*.

The *United States* has a welded steel hull but practically all of her interior work is aluminum. There is almost no wood in her. She is about as nearly fireproof as a ship can be made; even her paint is fire-resistant, and this is important because when a metal ship burns, the paint provides most of the fuel for the fire.

The Navy took an active part in the design of the new ship. As a result she will be quickly convertible into a troop transport, with a capacity of 14,000 men.

Aback: Square sails blown the wrong way on their yards are aback.

About: To go about, to turn the ship in tacking so as to bring the wind on her other side.

Accommodation Ladder: A set of steps for getting over the side of a ship from the water.

Aft (adverb): In, near, to, or towards the stern.

After (adjective): Hinder, as the afterdeck.

Aground: When a boat makes the discovery that all water has land under it, she is aground.

Aloft: In the rigging, above the deck.

Amidships: In the middle of the ship.

Anchor: Iron or iron-and-wood device to hold a ship in one place by digging into the sea bottom.

Arch Board: Early name for gunwale; from the medieval custom of supporting the fore and after castles on arches.

Astern: Behind the ship. To go astern is to move the ship backwards.

Athwartship: Across, in the direction of a thwart or rowing seat.

Avast!: Stop. Quit what you are doing.

Awash: Covered, but barely covered by water. As a derelict with her decks awash.

Aweather: To windward.

Back Ropes: Stays from the dolphin striker to the port and starboard bows.

Backing Link: A bent metal fitting fastened to the outside of the hull to which the lower end of a chain plate was secured.

Backstay: A rope led from the upper part of the mast back (and down) to ease the strain the wind puts on the mast. Reverse of forestay.

Bare Poles: With no sails set.

Beam: Width of a ship at the widest point. A point exactly abreast is "on the beam," a little ahead "before the beam," a little behind "abaft the beam."

Beam ends: Capsized. A vessel lying on her side in the water is on her beam ends, i.e., her deck beams are end down.

Beat: To tack to windward.

Becket Block: A block with an eye at one end for attaching the standing part of tackle.

Belay: To make fast (tie) a line, usually at the pin rail or at the fife rail.

Belaying Pins: Short rods set free in holes in the pin or fife rails on which a line can be quickly made fast and more quickly released by pulling the pin.

Bell: Time was struck on the ship's bell which hung in the belfry. Even numbers indicated hours, odd numbers half hours. So one bell is twelve-thirty, two bells one o'clock and so on, up to eight bells which is four o'clock. At that point the sequence begins again. One bell is four-thirty, two bells five, repeating again at eight bells or eight o'clock and again at eight bells, twelve o'clock. The modern navy has abandoned this in favor of a twenty-four-hour clock.

Bend: To fasten, as bending sail; also the name of a knot, the carrick (carrack) bend.

Bentick Shrouds: Lines once attached to the foot of the futtock shrouds and crossed over to tackle secured to a ring bolt in the deck by the opposite bulwark.

Bibbs (or hounds): Blocks fixed to the flattened sides of a masthead to support the trestle trees.

Bight: A loop of rope, whether secured by a knot, or a splice, or free.

Bilge: The curve where the side of the hull meets the bottom. Also the unpleasant liquid which accumulates inside below this point.

Billboards: Extra-heavy planking near the bow for the bill, at the tip of the anchor fluke, to rest against when the anchor is fished up and stowed against the ship's side.

Binnacle: Case or housing for the ship's compass.

Bitts: Heavy posts of wood or cast iron on the deck, to which lines are made fast. Most often they are in pairs. A single one is a bollard.

Block: Wood or metal case for sheave (pulley) or sheaves.

Boatswain (bos'n): The petty officer in charge of rigging and all deck gear.

Bobstay: A stay from the bowsprit tip to the stem almost at the waterline, pulls against the foretopmast stay.

Bolt Rope: A reinforcing rope sewn all the way around the edge of a sail.

Bonaventure Mizzen: A fourth mast used on Elizabethan and earlier ships.

Bonnet: A removable strip at the foot of a sail. Early method of reducing sail area in heavy weather.

Boom: A spar for extending the lower edge of a sail. Used on fore-and-aft sails and jibs; the jib boom.

Boom Irons: Iron fixtures on the yards to secure the studding sail booms.

Bow: The whole forward end of a boat.

Bowlines (bol'ns): Lines leading from the leech of a square sail, forward, for the purpose of pulling the leading edge closer to the wind. Bowline is also the name of a much used non-slipping, non-jamming knot.

Bowsprit: A spar projecting forward, originally only for rigging, such as forestay and bowlines. Later supported sails.

Braces: Ropes leading back from the ends of the yards to brace them, and the sail with them, against the wind.

Brail: To gather the sail loosely up to the yard by the buntlines. Lateen mizzen sails had lines on them called brails.

Breech: The inboard end of a gun. The term is used for muzzle-loading as well as breech-loading guns.

Breeching: Heavy rope from the bulwarks to the cascable of a truck gun, for checking the recoil.

Bulkhead: Any transverse partition on a ship, sail or steam, wooden or steel.

Bulwark: Solid rail along the ship's side.

Bumkin: A spar extended aft of the stern for securing the sheet of the spanker or driver; or a short strut on the quarters to receive running rigging.

Buntlines: Ropes operating from the yard to the lower edge of the sail for brailing it up and to help in handling the canvas.

Cabin: A space for passengers of rank. The cabin trunk was the projection of the cabin above the deck.

Cap: A block on the end of a spar with a hole for another spar to pass through it—at the masthead, on the end of the bowsprit.

Capstan: A revolving cylinder with a ratchet, driven by capstan bars, operated by a number of men, for the very heaviest pulling, as breaking out and raising the anchor.

Careen: To deliberately heel a ship over whether in the water (to reduce her draft and float her in shoal water) or ashore (in old days) to scrape her bottom clean.

Carronade: A short gun (about four feet long) taking a light powder charge for close-in fighting. First cast at Carron, Scotland.

Cartridge: A wool package of powder, shoved into the muzzle of a truck gun ahead of the wad and the shot.

Carvel-built: Method of planking in which the edge of each plank is set flush on the edge of the plank below it. Universal in large wooden ships.

Cascable: The ball cast on the rear end of a gun over which the bight of the breeching was placed.

Catheads: Short projections on each side of the bow for use in pulling the anchor aboard after it has been raised. Sometimes the anchor was carried against the ship's bow secured to the cathead or "catted."

Center Board: A movable board serving the purpose of keel on small vessels operating in shallow waters. It is lowered through a well in the center of the boat and can be raised to clear a shoal or to reduce drag when sailing before the wind.

Chain Plate: Short chain or iron rod bolted to the hull, extending upward to the channel and forming a loop around the lower deadeye. To take the pull of the shrouds.

Channel: Projecting "shelf" on the hull sides to spread the shrouds. Originally chain wale.

Charlie Noble: The galley (kitchen) smoke pipe.

Chase: To pursue, but also, guns for that procedure, as bow chase and stern chase.

Clew: The lower corners of a square sail or the after corner of a fore-and-aft sail.

Clew Garnet: Tackle to pull the lower corner of a square sail upward and inward on the after side.

Clinker-built: Method of planking in which the edge of each plank laps over the plank below it as clapboards are put on a house. Impractical for large boats.

Close-hauled: Sailing as nearly into the wind as possible, with all yards hauled up as far as they can be put.

Cockpit: A well, aft in a small boat from which the tiller and main sheet may be controlled. Often having seats for passengers.

Companionway: Steps between decks, usually very steep.

Compass Card: A circular card subdivided into "points" for determining direction by comparison with the constant needle. A "lubber line" is marked on the compass case to show the center line of the ship. To *box* the compass is to repeat the names of the points in order clear around the card as N., etc. The position of the lubber line to be maintained is called the compass bearing.

Course: The lowest sail on the masts' as the mains'l, the fores'l.

Crossjack (cro'jack): The lowest yard on the mizzenmast, on which a sail is rarely set. The French call it the "dry yard."

Crosstrees: Thwartship spreaders for the shrouds of the upper sections of the mast. They rest on the trestle trees which run fore and aft.

Crow's-foot. On early ships, the bridle into which important stays were divided.

Dahlgren Gun: Called soda bottle from its shape, invented just before the Civil War. The first major improvement in the way of accuracy and movability.

Davits: Metal or wooden derricks to raise boats from the water and swing them aboard.

Deadeyes: Pierced circular wooden blocks through which lanyards were rove for tightening the shrouds.

Deadrise: The angle of the hull from the keel to the turn of the bilge.

Decks: The "floors" of a ship. On a sailing three-decker they were the berth deck (lower), gun deck (middle), and the main (upper, or spar) deck.

Dolphin Striker: A spar extending downward from the bowsprit cap to spread the stays of the jib boom. Sometimes, especially on frigates, it was double.

Douse: To let go the halyards of a sail so as to stop it drawing instantly; an emergency maneuver.

Downhaul: Tackle for pulling stays'ls and jibs down the stays on which they are set.

Draft: The depth of water needed to float a boat clear of the bottom.

Drift: Move with tide or current; also move with wind to leeward, unintentionally.

Driver: Sometimes used as name for spanker, actually an extension of the spanker as a studding sail extends a square sail—or a large sail temporarily replacing the regular spanker.

Dunnage (once tunnage): Loose lumber used to wedge cargo, especially casks, in the hold of a vessel.

Ensign: The national flag of a ship flown at her stern from a staff on a steamer; from a halyard to the spanker gaff on a sailing vessel, or from the mizzen truck.

Entry Port: Opening in the side of a high-sided man-o'-war, for entering and leaving the ship.

Fake: One ring of a coil of rope, whether working, as on a capstan, or idle, as lying coiled on deck.

Falconet: A small swivel gun mounted on the bulwarks for repelling boarders.

Fathom: Six feet, a measurement of length.

Fid: A wooden pin to secure the squared end of the topmast or topgallant mast.

Fife Rail: A sturdy rail at the foot of a mast, having holes for belaying pins to make fast halyards and other running rigging.

Figurehead: Carved decoration at the tip of the head. Sometimes allegorical, sometimes a literal portrait.

Flare: The extension of the upper part of the hull over the lower, particularly at the bow. It appeared in clippers and packets and is usual in large steamships.

Flemish Horse: A short footrope at the outer end of the topmast yard.

Flush-decked: Having an unbroken deck from stem to stern.

Footropes: Lines hung below a yard for a man to stand on while working on the sail.

Fore-and-aft Rig: Having sails set behind the mast with spars parallel (when at rest) to the direction of the keel.

Forecastle: Raised part of deck in the bow.

Foremast: The first in line, a little shorter than the mainmast, a little taller than the mizzen.

Forestay: A stay leading from the foretop to the bowsprit cap and from there to the hull.

Freeboard: Height of the side of the hull from waterline to upper deck.

Funnel: The smokestack of a steamship.

Furl: To gather the sail tightly to its yard and "stop" it there with gaskets. To *hand* a sail is the same thing. Done only when the sail will not be needed for some time.

Futtock Shrouds: Short ropes, or rods, below the crosstrees to take the strain of the topmast shrouds. Until the nineteenth century they were secured to the lower shrouds, after that to a band around the mast.

Gaff: A spar for extending the peak of a fore-and-aft sail.

Galleries: Balconies across the stern and around the quarter of sailing ships in the sixteenth, seventeenth and early eighteenth centuries.

Galley: The kitchen on a boat. Also a large, oared craft.

Gammoning: Heavy lashings to hold down the bowsprit—later replaced with metal.

Gangway: Opening in the bulwarks for boarding or leaving the ship.

Gaskets: Bands of canvas or plaited hemp for binding the furled sails to the yards.

Go About: When "on the wind" to change from one tack to the other, to put the wind on the other side by heading into it and swinging over the other way.

Grating: A flat surface made of crossed and spaced strips of wood or of pierced metal. Used as hatch covers in good weather and as floor boards for small boats.

Gudgeons: The fittings on the sternpost to which the rudder is hung; the corresponding fittings on the rudder are pintles.

Gunport: Opening, usually square, in the hull or the bulwark to allow a gun muzzle to be run out.

Halyard: Tackle for raising sails and spars on the mast.

Hammock Nettings: Racks on top of the bulwarks of a wooden man-o'-war for stowing sleeping hammocks during action. This cleared the gun deck, which was sleeping quarters, and afforded additional protection to the gunners on the main deck.

Hatch: An opening in the deck, provided with a hatch cover and with a kind of box built around it to keep water out. This is called the hatch coaming.

Hawse Pipe: A hole in the hull at the bow for the hawser or chain attached to the anchor.

Head: A projection forward of the stem under the bowsprit, varying greatly in shape at different periods. Used formerly as the ship's toilet, which is still the "head."

Head and Heel: The head is the top or outboard end of anything; the heel, the bottom or inboard end. Applies to mast, bowsprit, rudder, etc.

Headsail: Any of the triangular sails set on the stays of the foremast. Headsails from foremast forward: 1. forestays'l, 2. foretopmast stays'l, 3. inner jib, 4 outer jib, 5. flying jib, 6. jib tops'l (jib-o'-jib).

Heeled Over: A ship leaning to one side because of pressure of wind on her sails. She is pretty well over with her "scuppers under" and very far over with her "rail under."

Horse: Another name for footrope. The short vertical lines which supported the footrope to the yard were called stirrups.

House Flag: The private flag of the owners flown from the foremast truck.

Hove-to: Stationary, head to the wind. Executed on a square-rigger by letting some sails draw and backing others.

Hull: The hull is the whole body of a boat. It is what floats.

Inboard: On the ship, inside the bulwarks or rail.

Jack: A national maritime flag flown at the bow (or foremast truck) of a ship. It usually differs in design from the national ensign. The American naval jack is a solid blue flag with a white star for each state. This is sometimes erroneously called the Union Jack which properly is the flag of the Union of Great Britain and Ireland.

Jackstay: Iron rod fixed to the top of a yard, for bending on sail. Formerly the sail was lashed to the yard as a whole.

Jacob's Ladder: A rope ladder up the after-side of the mast. Used on ships with tackle-tightened shrouds, where ratlines would not work.

Jamie Green: An extra sail used only on clippers. It was hung from the tip of the jib boom and set under it like a sprits'l, but was sheeted to one side only.

Jaws: The fork of a boom or gaff which allows it to swing around the mast.

Jeer: Tackle for raising and lowering the lower yards.

Jib: A headsail set forward of the forestays'l.

Jib Boom: A spar extending from the bowsprit to which the foot of the jib is secured. The flying jib boom is another spar extending from the jib boom.

Jibe: When sailing nearly with the wind, to change the direction of the ship so that the wind comes over the other rail. It is a violent and dangerous maneuver.

Jigger: A small sail aft on a fore-and-aft rigged boat. A yacht with mains'l and jigger is yawl-rigged.

Kedge (anchor): A light anchor with a movable stock. Also: to kedge, to move a ship ahead by pulling it up to an anchor with the capstan.

Keel: The long timber from end to end of the outside of a ship's bottom. The keel keeps the vessel from being blown sideways, making leeway.

Keelson: A long timber from end to end of the inside of a ship's bottom. It is above the keel.

Knee: A kind of bracket attached to a rib and helping to support a deck beam. They are also similarly used elsewhere.

Knightheads: Vertical posts between which the heel of the bowsprit was fitted. Later a single post with a hole in it was used.

Lanyard: A short, light line used as a lashing.

Lateen Sail: A triangular sail bent to a long yard.

Lead: If your pronounce it *leed*, it is the direction a line takes from where it acts to where it is belayed. If you pronounce it *led*, it is the weight to sink the lead line for measuring depth of water.

Lee: The side towards which the wind is blowing, the opposite of the windward or weather side. To put the helm alee is to turn the ship down wind.

Leeboards: Boards swung over the sides to act like a keel in preventing leeway. They can be raised like a center board. Used in small vessels, especially in Holland and on the Thames in England.

Leech: The side edge of a square sail and the after edge of a fore-and-aft sail.

Leech Line: Similar in purpose to a buntline but attached to the side of the sail.

Line: The sailor's word for a rope.

Lubber's Hole: Opening in the top at the head of the shrouds to permit easy access to the top, supposed never to be used by a sailor. A lubber is a sailor ignorant of his work.

Luff: The forward edge of a fore-and-aft sail. Also that side of a square sail set nearest the wind. To luff is to bring the ship up so nearly into the wind that the luff begins to flutter.

Mainmast: Just that. Second from the bow whether there are two or more. Almost invariably larger and taller than any other. Exceptions are yawls and ketches where the forward mast is the main.

Man (verb): Prepare to act, as "man the yards" for handling sail, or "man the guns."

Marconi Rig: A modern fore-and-aft rig which features a triangular mains'l and (if any) jigger, eliminating the heavy gaff and the tops'l. It is easily handled by a short-handed crew. Very popular on sailing yachts.

Marlin or Marline: Light tarred twine used for serving and seizing. In the eighteenth century and before, something called spun yarn was used. This was fiber raveled out of old rope. Oakum was hemp or jute fiber soaked in tar and wedged between the planks of a wooden ship to caulk it, or make it watertight.

Marlinspike (also fid): A tapered spike usually about a foot long used to open the strands of rope for splicing, but also serving handily on occasion as a weapon for private combat.

Martingale: Stay from the dolphin striker to the head of the jib boom, pulling against the topgallant stay.

Mast: A vertical "stick" in a boat for supporting sails and rigging. In large craft usually in three parts, overlapped and secured together.

Mizzen (mast): The third mast. The last mast back. The smallest and shortest.

Monkey Rail: A short pin rail in the bow for belaying headsail downhauls.

Mortar: A very short gun, built to throw a large shot at a very high angle.

Muzzle: The outboard end of a gun.

Offing: Seaward, away from shore.

Orlop Deck: The first deck above the bilge. Usually it had little head-room (four feet) and was used for storage.

Outboard: On the hull outside the bulwarks or rail.

Outhaul: Rope gear on both sides of a gun, attached to the rear of the carriage and to the bulwarks, for running the gun out to fire.

Painter: A line in the bow of a small boat (usually made fast there) for tying her to ship, mooring, or landing stage.

Parrel: A sort of beaded collar around the mast for holding yards on early sailing ships, and booms and gaffs on later ones.

Partners: Extra-heavy timbers in the deck where the mast passes through it.

Pay (verb): To paint with tar for waterproofing. Also to to let a line run—"pay out a line."

Peak: The head of a gaff, or the upper after corner of a fore-and-aft sail, attached to the gaff at that point.

Pendant: A short line to hold a block (pulley).

Pin Rail: A sturdy rail along the bulwark, with holes for belaying pins.

Pitch: The angle of a mast leaning forward. The opposite is rake.

Pole: The extreme upper end of the mast, or a simple one-piece mast without fittings for a topmast.

Poop: The high deck at the stern. The name comes from a similar deck on Roman ships where the images of the gods, the *pupae,* were kept.

Pooped: Overtaken by a following wave.

Port: The left side of a boat, as the steersman stands, facing forward.

Preventer: The temporary replacement of a broken line, or an auxiliary to take over if the main line breaks.

Pump: A permanent fixture, nearly always abaft the main-mast, for the daily removal of accumulated bilge water.

Quarter: The outside of the ship between the main chains and the stern.

Quarter Bumkin: A strut or short spar on the quarter with a block to receive the main brace.

Quarter-deck: The next lower section of deck forward of the poop. In later naval vessels the poop was left off and the quarter-deck ran all the way to the stern.

Quoin: A large wooden wedge shoved under the breech of a truck gun to lower the muzzle. The breech was levered up with a crowbar (since the gun weighed 5,550 pounds) and the quoin moved forward or back as needed.

Rabbet: The groove along the keel into which the garboard strake is fitted.

Rail Cap: The rounded top of the bulwarks.

Rake: The angle of a mast leaning backward. The opposite of pitch.

Raked: A sailing ship was raked when she took a broadside into her stern, or worse, her bow, which was almost defenseless.

Ram: Projection from the bow near water level for damaging other vessels. Used on early Mediterranean galleys and on battleships during and after the Spanish-American War.

Rammer (ramrod): A long-handled implement for pushing powder and shot into the muzzle of a truck gun.

Rates: In the British Navy, size, based on the number of guns carried. Ships of the line: First Rate, one hundred guns; Second Rate, ninety to ninety-eight guns, three gun decks; Third Rate, sixty-four to eighty-four guns, two gun decks. Ship: Fourth Rate, fifty guns, two gun decks. Frigate: Fifth Rate, thirty-two to forty-four guns; Sixth Rate, twenty to twenty-eight guns. Sloop: (no rate) fourteen to eighteen guns, regardless of rig or number of masts.

Ratlines: Light tarred lines secured horizontally across the shrouds at regular intervals to serve as ladders into the rigging.

Rattle: A hand-operated signaling device, used on old fighting ships to command attention in the noise of battle.

Reciprocating Engine (steam): An engine which develops power from the out and back movement of a piston in a cylinder. In the days of low pressure, steam cylinders ran around six feet in diameter and the piston moved about four feet.

Reef Points: Short, light lines secured through a sail and hanging down on both sides, for "reefing" the sail to smaller size in high winds, by tying it up to its yard if a square sail, or down to its boom if a fore-and-aft sail. Headsails are reefed to bowsprit or jib boom.

Reeve (past tense "rove")· To pass a line through a hole, a grommet, or a block.

Ribs: Curved wooden frames, usually fastened to the keel, to which the planking of the hull is secured. Sometimes called timbers.

Rigging: The cordage and spars necessary to hold the sails in place and manage them. Often called tophamper.

Right (verb): A ship *rights* herself when she returns to normal after heeling over; a helmsman *rights* the helm when he lines the rudder up with the keel.

Robands (rope bands): Short lashings at the head of a sail to fasten it to the yard.

Royal (never royal sail): The sail above the topgallants'l. Set on an extension of the "gallant" mast. The name came from the trick of setting a sail on the flagpole used for the Royal Standard.

Rubbing Strake: A wale at the point of greatest width of the hull to take the rub of wharves and of other boats.

Running Rigging: Cordage for moving sails as: braces, sheets, tacks, bowlines, halyards and downhauls.

Saddle: A wooden cleat on an upper (small) yard where it rests against the mast.

Sails: Are named from the mast which supports them (fore, main, mizzen) and from their position on the mast. The lowest uses only the name of the mast, thus, from the deck upward there is the fores'l, foretops'l, foretopgallants'l, foreroyal and (rarely) the foreskys'l, and so on for each mast. See diagram facing title page.

Screw: The propeller of a steam- or motor-driven vessel.

Scud: To run before the wind, especially under shortened sail or bare poles, when the force of the wind allows no other maneuver.

Scupper: A hole in the bulwark at deck level to allow water to run off.

Scuttle: To sink a ship deliberately, by cutting a hole in her bottom, or by opening her sea cocks.

Seize and *Stop*: To lash two or more pieces of rope together is to seize them. To lash a loose end to its own standing part is to stop it.

Serve: To wrap a rope with marlin or other light cordage in order to stop fraying or reduce wear.

Sheave: The grooved pulley in a block.

Sheer: The curve or dip of the upper part of the hull from bow to stern.

Sheets: Lines fastened to the clews (lower corners) of a square sail or to the lower after corner of a fore-and-aft sail to hold against the pull of the wind.

Shin-cracker: A steering wheel mounted on and traveling with the tiller, operated by tackles secured in the port and starboard scuppers.

Ship: At first any large vessel. Later it described the three-masted square-rig, as well.

Ship (verb): To join a ship as a sailor. Also, to take sea water over the side, as to ship a sea (wave).

Ship of the Line: A ship large and powerful enough to be used in the first line of battle.

Shortening Sail: Reducing the area, either by reefing, or by furling some of the sails completely.

Shrouds: Heavy lines from the mast to the sides of the ship, to support the mast athwartship.

Shutter: The cover of a gunport. On lower decks of one piece, lifted by a lanyard; on the upper deck, made in two halves, the lower half falling to the horizontal, the upper removed entirely.

Side-wheeler: A steamboat driven by paddle wheels at the sides. A stern-wheeler has only one paddle wheel, aft. Both are paddle boats.

Skids: Vertical timbers set outside the planking and the wales. Their chief use was in canting a vessel on shore for scraping her bottom.

Skys'l: A sail set above the royal. Very rarely moon-rakers have been set above skys'ls.

Slings: The old way of hanging a yard on the mast; on later ships iron fittings called trusses or cranse irons were used.

Spanker: A fore-and-aft sail on the mizzenmast of a square-rigged ship.

Spar: Any support for sails or rigging as mast, yard, boom, gaff, sprit, and bumkin.

Spar Deck: The upper deck of a sailing ship.

Spencer: An extra fore-and-aft sail, set aft of the main-mast on a spencer mast, parallel to the mainmast.

Spill: That is, spill the wind from a sail by slacking the braces, so the sail may be handled by men on the yardarms.

Splice: To join the ends of two ropes by interweaving the strands of each into the other. A short splice is a quick job which bulges the rope too much to pass through a block. A good long splice will pass and increases the thickness of the line very little at the joint.

Sponge: A sheepskin-covered ramrod, used wet, to clean a gun barrel after firing.

Spring Stay: A stay set horizontally between mastheads.

Sprit: A spar which crosses the middle of a sail in extending it.

Sprits'l: A sail hung from the bowsprit, or a sail extended by a diagonal, or a horizontal sprit.

Sprits'l Tops'l: A small, square sail set on a short mast at the bowsprit tip, or on a spar projecting like a jib boom from the bowsprit.

Square-rig: Having sails set in front of the mast with spars (when at rest) lying across the direction of the keel.

Stanchion: An upright support for a rail or awning.

Stand By: Prepare: to receive orders, to board, to repel boarders, for grub, for grog, for anything.

Standing Rigging: Fixed cordage, as: shrouds, forestays and backstays used to support and strengthen the masts.

Starboard: The right side of a boat as the steersman stands, facing forward. From steerboard side, the Scandinavian name for the side where the rudder was hung.

Stay: A rope led from the upper part of a mast forward and down, pulling against the backstays. The stays on the foremast are led to the bowsprit assembly. Those on the main to various points on the foremast, those on the mizzen to various points on the main. They are named for the part of the mast to which their higher ends are attached; as: the main stay, main topmast stay, main topgallant stay, main royal stay.

Stays'ls: More or less triangular fore-and-aft sails set on the stays ahead of the masts of square-rigged ships.

Steeve: The upward angle of the bowsprit, measured from the horizontal.

Stem: The upright timber at the forward end of the keel, to which the ends of the planking are brought.

Step: A pierced block on the keelson which receives the foot of a mast.

Stern: The whole after or back end of a boat.

Sternpost: The more or less vertical timber at the after end of the keel.

Stock: The crossbar of an anchor, once made of wood.

Stow: To put away in its proper place; applied to anything loose on one end or both.

Strake: One line of planking on a wooden ship. One line of plates on an iron ship. The garboard strake is at the keel, the bilge strake is at the turn of the bilge, the sheer strake at the top of the bulwarks under the rail cap.

Studding Sails (stuns'ls): Extra sails, set on booms extending from the ends of the yards, for use in light air. Normally set on the foremast and mainmast only and seldom higher than the topgallants'l.

Swiftering: Zigzag lacing together of the two parts of a double stay.

Swivel Gun: A small gun readily moved by a sort of handle at its back end. Firing scrap iron (canister) and used to repel boarders. Usually mounted on the bulwarks.

Tack: The line that holds the clew of a square sail forward when sailing "on the wind," hence to tack is to sail into the wind. Tack is also the bottom forward corner of a fore-and-aft sail.

Tack (verb): To progress against the wind by sailing a zigzag course, as nearly into it as possible. The starboard tack has the wind on that side, and the ship is headed to the left of her actual destination. The port tack reverses this.

Tackle (formerly pronounced Taykle): Rope rigged between and around pulleys to increase the effect of pull applied.

Tafferel (not taffrail): The extreme top of the stern of a ship, where the old Dutch had a picture, now wrongly used to designate the after rail.

Throat: The forward upper corner of a fore-and-aft sail.

Tiller: A lever for operating a rudder.

Timberheads: On wooden ships the projection of the frames (ribs) above the deck, to support the bulwarks.

Toggle: A wooden pin used to secure the blocks for the sheet, tack and clew garnet to the lower corner of a square sail.

Tonnage: Not the weight but the carrying capacity of a ship, originally tunnage, that is, the number of tuns (casks) of wine she could carry. Tonnage is arrived at by complex formulas of measurement which have varied from time to time and from nation to nation.

Top: A platform, at the lower crosstrees; from the medieval fighting top.

Topgallants'l: The third sail from the deck (when the tops'l is single). The sail above the tops'l; set on the topgallant mast. Usually shortened to gans'l.

Topping Lift: A rig to raise and control the vertical angle of a yard or boom.

Tops'l (square): The second sail from the deck, set on the topmast above the top. It became so large and unwieldy that it was divided into two sails in the nineteenth century; called upper and lower tops'l.

Tops'l (Fore-and-Aft): A sail rigged between the gaff and the topmast. Usually it is triangular or triangular with the lower corner cut off.

Train Tackle: Rope gear hitched to the rear of a gun carriage for holding it inboard after firing, for cleaning and reloading.

Transom: The flat part of the stern in a square-sterned vessel.

Traveler: A thwartship rod a few inches above the deck, fitted with a slider to which is caught the sheet of a fore-and-aft sail.

Truck: The cap or block at the extreme top of the mast, usually fitted for flag halyards. The wheels of a gun carriage were also called trucks.

Truck Gun: The standard eighteenth-century naval gun mounted on trucks, or wheels. One firing a thirty-two-pound round shot would be nine feet, six inches long, with six-and-a-half-inch bore and would throw the shot about twelve hundred yards.

Trunnions: Cylindrical lugs cast on the side of a truck gun to support it on the gun carriage.

Truss: On later sailing ships trusses were iron fittings to hold a large lower yard on the mast.

Tumble Home: The slope inward of a ship's sides; having the hull narrower at the main deck than at the waterline.

Turbine: The modern steam engine for large ships. Steam is converted directly into rotating power by acting against hundreds of small vanes set in a large, enclosed rotor.

'Tween Decks: Anywhere in the body of the ship, above the bilge and below the spar deck.

Tye: That end of a halyard which is attached to the yard. Early tyes were of rawhide, later ones of chain.

Vang: Line from a gaff to the rail to steady the gaff—fore-and-aft sail.

Waist: The lower part of the deck, between the quarterdeck and the forecastle.

Wale: Long timbers outside the planking of the hull to

protect the ship in collision and when "canted" ashore for bottom-scraping. Only the gunwale survives. Pronounced gunnel. There were also the main wale and the bilge wale. Sailors often called them bends.

Walking Beam: A diamond-shaped rocker arm above the upper deck of a paddle steamer, to transfer power from the engine to the paddles.

Ware Around: When on the wind, to change the tack by going down wind and executing a controlled jibe. A tricky stunt not often executed. Also called boxhauling.

Waterways: Gutters along the bulwarks to lead water to the scuppers.

Wedges: Usually eight, set into the deck around the mast to hold it very tightly.

Wheel: The propeller of a steamship. The steering wheel of a ship, sail or steam. The steering wheel was not known before 1711.

Whipstaff: Steering device consisting of a vertical lever for moving the tiller. Obsolete after 1711.

Whisker Boom: Horizontal spar at the bowsprit head to spread the stays of the jib boom. It took over the job of the earlier sprits'l yard.

Windlass: A horizontal hand winch, crank operated, for extra heavy lifting.

Windward and Loo'ard (leeward): The windward side of a ship is the side against which the wind is blowing. It is not good manners at sea to spit to windward. The opposite side is the lee or loo'ard side. A lee shore is one that lies over the lee rail, dangerous because the wind tries to blow the ship against it.

Wolding: Originally rope wrapped in bands to strengthen the lower mast. Later iron was used.

Worm (verb): To lay light line in the grooves between the strands of a large rope before serving, so as to achieve a smooth, cylindrical finished job.

Yards: The horizontal poles which cross the mast. They support the sails. The two ends are yardarms but *the* yardarm, which was used as a gallows, was usually the crossjack of the mizzenmast.

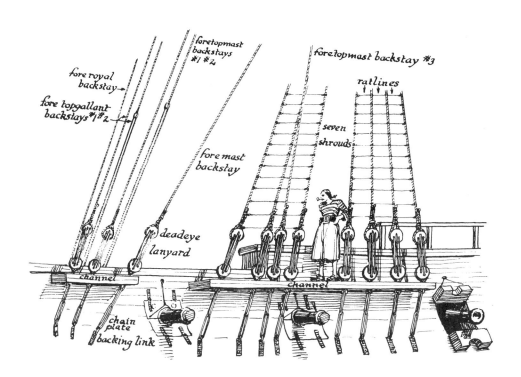

Channel Rigging of the Frigate Raleigh

Eighteenth-century Marblehead Heeltapper

with Sails Reefed